I0055476

Small-Aperture
Radio Direction-Finding

For a complete listing of the *Artech House Radar Library*,
turn to the back of this book . . .

Small-Aperture
Radio Direction-Finding

Herndon H. Jenkins

Artech House
Boston • London

Library of Congress Cataloging-in-Publication Data

Jenkins, Herndon H.
    Small-aperture radio direction-finding / Herndon H. Jenkins.
       p.   cm.
    Includes bibliographical references and index.
    ISBN 0-89006-420-2
    1. Radio direction finders. I. Title.
  TK6565.D5J46 1991   91-9170
  621.384'191--dc20   CIP

British Library Cataloguing in Publication Data

Jenkins, Herndon H.
    Small-aperture direction-finding.
    1. Radio frequency equipment
    I. Title.
    621.384191

    ISBN 0-89006-420-2

© 1991 Artech House, Inc.
685 Canton Street
Norwood, MA 02062

All rights reserved. Printed and bound in the United States of America. No part of this publication may be reproduced or utilized in any form or by any means, electronic or mechanical, including photocopying, recording, or by any information storage and retrieval system, without permission in writing from the publisher.

International Standard Book Number: 0-89006-420-2
Library of Congress Catalog Card Number: 91-9170

10   9   8   7   6   5   4   3   2   1

Contents

Preface

Since its inception near the turn of the twentieth century, the area of small-aperture radio direction finding (DF) has grown and evolved. Today, small-aperture direction finding plays a major role in the broader areas of passive radiolocation and radionavigation. Numerous techniques have been formulated, and many operational systems have been developed to satisfy a wide variety of technical and operational requirements. For example, small-aperture DF systems are used for military, paramilitary, government, public service, commercial, and civilian applications.

The primary purposes of this book are to delineate the character and attributes of small-aperture radio direction finding and to present a unified approach to the system-level aspects. The emphasis is on fundamental principles of operation and basic performance capabilities. Major objectives are to identify error-producing mechanisms, quantitatively define parameters, and present error-mitigation practices. To enhance its utility, this book includes material on subsystem considerations, test and calibration, DF networking, and passive geolocation by radio direction finding.

This book should be of interest to a wide audience. It can serve as an introductory text for the entry-level DF practitioner and as a technical reference for the experienced DF specialist. The book should be of value to both engineers engaged in DF research, design, and development and those responsible for operational deployment and use. System engineers engaged in integrating small-aperture DF systems into larger systems will find this volume useful. Similarly, small-aperture DF users and operators can benefit from it.

Analytical results are presented in a form convenient for engineering use. A knowledge of mathematics through calculus is desirable. Rigorous mathematical treatments have been deemphasized in the interest of brevity and conciseness; however, sufficient mathematical detail is given to support the analytical results, and, where possible, references are made to sources containing more detailed mathematical treatments.

Chapter 1 is concerned with the functional elements of radio direction-finding and a definition of small-aperture DF based on linear wavelength criteria. History and evolution are discussed. Chapter 2 presents the basic operating principles of radio direction finding, in general, and small-aperture DF, in particular. Signal direction-of-arrival measurement techniques using amplitude, phase, and time are introduced. Technical and operational performance factors in the frequency, space, and time domains are discussed in Chapter 3. Chapter 4 is concerned with direction-of-arrival error sources including propagation-induced errors and environmental, instrumental, and observational errors. The effects on small-aperture DF performance are stressed. In Chapter 5, small-aperture DF techniques are categorized based on the method of deriving the direction-of-arrival information. Each technique is analyzed, and the method of operation is quantitatively defined. Strengths and limitations are delineated. Chapter 6 presents representative examples of contemporary operational small-aperture DF systems. Chapter 7 deals with passive geolocation techniques including DF homing, navigation, and triangulation. Subsystem characteristics and design approaches that significantly impact performance are described in Chapter 8. In Chapter 9, DF calibration and test are discussed. Chapter 10 includes an analysis of direction finding on frequency-hopping, spread-spectrum transmissions. The appendix describes method of moments computer programs suitable for small-aperture DF analysis.

The reference material for this book was generated during my work on small-aperture RDF while I was employed by the Georgia Tech Research Institute (GTRI) of the Georgia Institute of Technology. Preparation of the book was enhanced by a long association with my colleagues at GTRI especially Messrs. Richard Moss, Bruce Warren, and Bobby Wilson. The support and contributions of Mr. Lawrence Scott of the U.S. Army Electronics Research and Development Command are gratefully acknowledged. I am very appreciative of the technical critique performed by an anonymous reviewer. Special thanks are due to Ms. Pat Salomon for expertly preparing the illustrations. The patience and encouragement of my family and friends were invaluable.

Chapter 1
INTRODUCTION

1.1 DEFINITIONS

Radio direction finding is a class of direction finding by which the direction to a radio source is determined by means of a radio direction finder (RDF). A radio direction finder is a passive device that determines the direction of arrival (DOA) of radio-frequency energy. An RDF is a receiving system that operates on the energy extracted from the passing electromagnetic radio wave to obtain DOA information [1].

A radio direction finder consists of four essential functional elements as depicted in Figure 1.1. The antenna extracts electromagnetic energy and converts it to a signal containing direction-of-arrival information. The receiver converts, amplifies, and processes the signal to IF or baseband. The postreceiver processor further processes the signal to obtain basic angle-of-arrival (AOA) information. The DF information processing–read-out–display unit prepares the basic AOA data for transmission to users of the DF information.

The antenna is the key subsystem in an RDF. It extracts energy from the incident electromagnetic field and provides output signals containing incident energy AOA information. Radio direction finders are categorized on the basis of directivity, which is a measure of the antenna's ability to receive energy selectively in the space domain. The most common measure of directivity is half-power beamwidth, which can, in turn, be related to the linear aperture of the antenna. (The linear aperture is defined as the largest linear dimension of the antenna.)

Figure 1.2 shows half-power beamwidth plotted *versus* linear aperture in wavelengths. Regions are delineated for small-aperture, medium-aperture, and large-aperture RDFs. The bounds between aperture classes are not rigorous. (Sometimes narrow-aperture is used for small-aperture and wide-aperture is used for large-aperture. No standard exists; however, common usage favors small-aperture and large-aperture RDF). Figure 1.2 also shows representative direction-finding antenna types categorized by linear wavelength.

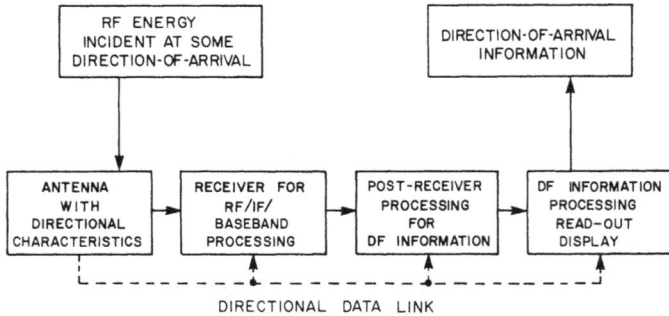

Figure 1.1 Functional elements of the direction-finding process (after Bailey [2]).

The delineation between the small-aperture and medium-aperture regions is not well defined. Some DF practitioners state that the small-aperture region is for linear apertures that are a small fraction of a wavelength; other practitioners state that the small-aperture region extends to half-wavelength apertures. This book will adhere to the half-wavelength criteria because the current trend in broad-band-width, small-aperture DF is to use an aperture at or near a half-wavelength at the highest operating frequency.

The "optimum" design of a small-aperture DF involves a unique set of design considerations and parameter trade-offs. The deleterious effects of a low antenna

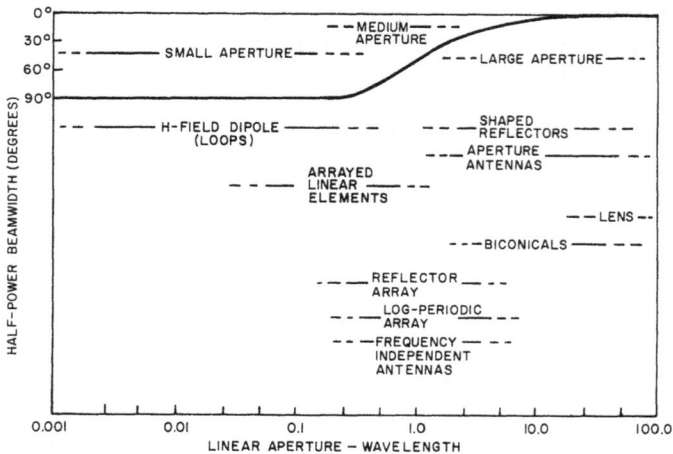

Figure 1.2 Beamwidth *versus* linear aperture in wavelength with direction-finding aperture regions and representative antenna types (after Reich [1]).

effective area, circuitry amplitude and phase unbalance, antenna element coupling and scattering, and incident signal depolarization and multipath are exacerbated for small-aperture usage. Obviously, the use of a medium or large aperture is preferred; however, the required operating range, limited physical space, and the demands of the host platform siting may require the use of a small-aperture DF. Recently, other operational demands, such as visual and radar detection covertness, have helped increase the emphasis on small-aperture DF use.

The required operating frequency range and the physical size of the antenna aperture that is practicable to build usually determine the type of DF: small, medium, or large aperture.

1.2 HISTORY AND EVOLUTION

Small-aperture radio direction finding, which is a constantly evolving discipline, began during the early years of the twentieth century as a direct result of work performed by Hertz, Marconi, and Zenneck on directive antennas [3]. In 1899, Brown rotated a pair of interconnected vertical antennas spaced a half-wavelength apart to obtain directive radiation [4]. In 1902, Stone proposed [3] the rotation of the array developed by Brown [4] to acquire direction-of-arrival information. In 1906, Marconi performed direction-finding experiments on signals transmitted from the *H.M.S. Furious,* which was steaming 16 miles off shore. Marconi used a radial system of horizontal antennas about one-fifth of a wavelength long, suspended close to the ground. The experiments by Brown and Marconi were probably the first systematic use of radio direction finding.

The advantages of a vertical loop DF antenna were recognized in 1907 as Pickard and DeForest concentrated on loop antenna investigations [4]. Pickard conducted loop antenna direction-finding field tests that established the vertical loop antenna as the dominant DF antenna.

In 1903, Bellini and Tosi developed the radio goniometer [3, 4] and applied it to two crossed loops to attain the first electromechanically scanned small-aperture direction finder.

Experience with loop antenna direction finders quickly disclosed that they are subject to large errors in the presence of ionospherically propagated signals with horizontal polarization. In 1918, the Adcock antenna was developed, although Stone appears to have had the initial concept of the Adcock [3]. Consisting of two vertical antennas (dipoles or monopoles) connected in phase opposition, the Adcock significantly reduced the error created by horizontally polarized, sky wave signals.

During World War I, the vacuum tube receiver was developed. Its use greatly improved the performance of radio direction finding.

Introduced in 1926, the next major small-aperture direction-finding technique [4] was the Watson-Watt DF, which provided instantaneous angle-of-arrival information. Two crossed, quadrature loops have their outputs applied directly to opposite plates of a cathode-ray oscillograph. This provides for an instantaneous sum of the two quadrature sine pattern functions of the two loops. In 1934, the technique was applied to Adcock arrays. Twin-channel amplitude- and phase-balanced receivers were incorporated in 1938 to improve performance significantly and provide the first truly instantaneous direction finder.

By 1940, the first generation of small-aperture direction finders was in operation—primarily in the 0.1- to 30-MHz band. During the first generation period, technique and hardware developments were paralleled by ionospheric propagation research. Emphasis was placed on investigating errors introduced into small-aperture direction finders by ionospherically propagated signals with horizontal polarization, combination ground and sky wave modes, multipath interference, sky wave lateral deviations, and traveling ionospheric irregularities. In addition, considerable analytical and experimental work was performed on errors introduced by nonoptimal siting conditions. Numerous error mitigation methods were attempted; many produced only limited success.

Variants of the basic simple loop, crossed loop, Adcock, Bellini-Tosi, and Watson-Watt implementations emerged. For example, a spaced-coaxial loop variant of the Adcock was developed and proved to be effective in reducing sky wave errors. Other significant improvements were (1) techniques to resolve the 180° angle-of-arrival ambiguity inherent in all the basic techniques and (2) mechanical commutation between multielement DF arrays to reduce bearing acquisition time.

In 1920, the U.S. National Bureau of Standards (NBS) established a network of radiobeacons operating in the 285- to 315-kHz band for marine radio direction finding. NBS also developed the loop direction finder for use with the radiobeacon system. Work on automatic RDFs accelerated and, by the late 1930s, they were widely used for both air and marine navigation. (A radiobeacon network for air navigation was deployed in the United States in the 1930s.)

World War II brought rapid improvements in and created widespread use of military versions of small-aperture direction finders, including VHF and UHF implementations. Operation of these systems under a wide variety of conditions disclosed not only the military advantages but also the inherent limitations. These limitations were extensively documented and helped form the rationale for concentrating on medium- and large-aperture direction finding after World War II.

During the war, RDF networks were widely employed. A notable example was the U.S. Navy DF net that located hostile submarines using HF transmissions. These nets, requiring relatively high levels of communication, command, and control, were the first application of DF usage as integrated, dispersed sensors. Also, during World War II, DF homing techniques were implemented and extensively used by aircraft, boats, and personnel.

In the mid-1950s, research and development on small-aperture DF increased. Emphasis was placed on defining the fundamental limitations on performance imposed by the ionosphere [5]. Exploratory systems were tested with ionosounder data. Other investigations concentrated on new configurations of conventional small-aperture DFs under severe shipborne siting conditions [6]. Error-reduction techniques for the simple loop DF were developed and tested. In general, these research and development efforts enhanced knowledge of small-aperture DF error sources and fundamental limitations. Some error mitigation techniques emerged; however, many error sources, such as operation in the presence of combined ground and sky wave interference, continued as major problems for small-aperture DF.

During the Vietnam conflict, hostile forces used horizontally polarized, near-vertical incidence HF ionospheric propagation for communication. Near-vertical incidence signaling (NVIS) produces horizontally polarized signals that compromise small-aperture DF performance. Attempts to use dual, orthogonal baseline, crossed loop arrays on NVIS signals met with limited success.

Burst communication modes, used by hostile forces during the Vietnam conflict, stressed small-aperture, noninstantaneous DF performance. The need for instantaneous DF or rapid electronic scanning was clearly evident and motivated upgrading techniques. The advent of HF and VHF spread-spectrum modulations has further increased the need for instantaneous or near-instantaneous DF. Angle-of-arrival information has emerged as a major discriminant for spread-spectrum signal sorting.

The last few decades have seen major improvements in small-aperture DF design and development. A second generation has evolved based primarily on phase interferometry and time difference-of-arrival techniques. In general, DF systems based on amplitude-only DF are used with cooperative emitters such as radiobeacons for homing and navigation, emergency locator transmitters, wildlife tracking sources, and personnel or vehicle transmitters. Automatic DF (ADF) systems, designed for homing and navigation, have incorporated advanced amplitude-null servo techniques and expanded coverage to the VHF region.

Second-generation small-aperture DFs for uncooperative emitters have broadband, instantaneous, or near-instantaneous performance with advanced signal processing algorithms for AOA error reduction. Netted operation provides rapid signal set-on and enhanced computer support. Real-time ionosounder data are often used to correct or compensate for errors introduced by ionospheric propagation.

Developments in the following areas have significantly enhanced phase and differential time-of-arrival techniques for small-aperture applications:

1. Broadband antennas;
2. Active antennas;

3. Low-noise, broadband RF preamplifiers;
4. Electronic antenna scan;
5. Broadband RF components;
6. Inherent sense resolution;
7. Balanced transmission lines;
8. Matched amplitude and phase, multichannel receivers;
9. Synthesized receivers, with digital tuning;
10. Advanced IF–video–digital, signal processing;
11. Microprocessors;
12. Robust angle-of-arrival algorithms;
13. Tactical computers;
14. Precision phase comparators, time interval measurements;
15. Real-time ionosounder support;
16. Rapid-erectible, free-standing antenna masts;
17. Interactive DF networks.

Currently, a considerable amount of research and development is being devoted to HF single station location (SSL). Medium-aperture versions are gaining operational status; small-aperture implementations will surely follow. Direction-of-arrival information on ionospherically propagated signals, combined with iono-sounder data, can be used to locate the position of the signal source by ray-tracing techniques. The attainable accuracy is a function of the quality and quantity of the ionosounder data.

To summarize, small-aperture direction finders have steadily evolved over nine decades. First-generation systems, using amplitude and amplitude-to-phase techniques for DF information, have been used effectively for numerous applications. However, major performance limitations exist when used on ionospherically propagated signals. Second-generation systems have concentrated on phase and differential time-of-arrival techniques and have benefited from advances in receivers, microprocessors, components, and devices.

In the future, small-aperture DF techniques may be required in frequency bands above the VHF range as the size of the host platforms decreases and the demand for covertness increases. Also, the structure of modern uncooperative emissions will be a major driver in establishing future small-aperture DF architectures.

1.3 MODERN APPLICATION AREAS

Small-aperture direction finding has found widespread use in numerous application areas. Listed below are some of the major areas with representative functions delineated by cooperative and uncooperative emitter usage.

- *Military:*
  - Cooperative Emitters
    Air and marine navigation
    Search and rescue
    Para-rescue
    Landing–drop-zone location
    Drop-zone assembly
    Personnel and vehicle locators
    Emission control
    Frequency management
  - Uncooperative Emitters
    Communication intelligence (COMINT)
    Electronic order of battle (EOB)
    Electronic support measures (ESM)
    Emitter homing and targeting
    Friendly force location
    Force strength assessments
    Interference source location
- *Civilian:*
  - Cooperative Emitters
    Air and marine navigation
    Emergency beacon location
    Search and rescue
    Para-rescue
    Wildlife tracking
    Personnel and vehicle location
    Position radio markers
  - Uncooperative Emitters
    Spectrum monitoring
    Spectrum density calibrations
    Clandestine transmitter location
    Lightning sferics location
    Amateur radio-frequency management
- *Government:*
  - Cooperative Emitters
    (same as civilian applications listed above)
  - Uncooperative Emitters
    Regulatory enforcement
    Spectrum density calibrations
    Spectrum monitoring
    Para-military

- *Research:*
  - Uncooperative Emitters
  Advanced COMINT investigations
  DF network architectures
  Radio noise definitions
  Electromagnetic studies of severe weather
  - Cooperative Emitters
  Advanced modulation DF vulnerability
  Propagation studies
  Component and device evaluation
  Remote environmental sensing

REFERENCES

1. Reich, H.J., ed., *VHF Techniques,* Vol. 1, New York: McGraw-Hill Book Co., 1947, pp. 223–225.
2. Bailey, A.D., "HF Direction of Arrival Studies Over a Medium Range Path," *Proc. Conf. HF Radio Propagation,* Urbana, IL, August 1970, pp. 13–30.
3. Boyd, J.A., D.B. Harris, D.D. King, and H. W. Welch, Jr., ed., *Electronic Countermeasures,* Los Altos, CA: Peninsula Publishing: 1978, Chapter 10: "Direction Finding" by L.A. deRosa, pp. 10-3–10-4.
4. Travers, D.N., and S.M. Hixon, "Abstracts of the Available Literature on Radio Direction Finding: 1899–1965," San Antonio, TX: Southwest Research Institute, Contracts NObsr-64585/85086/89167, July 1966, pp. 1, 3, 37.
5. Bailey, A.D., J.D. Dyson, and E.C. Hayden, "Studies and Investigations Leading to the Design of a Radio Direction Finder System for the MF-HF-VHF Range," Urbana, IL: University of Illinois, Electrical Engineering Research Laboratory, Final Report No. 20, Project No. DA 36-039-AMC-03720(E), ASTIA No. AD 453 179, July 1964.
6. Travers, D.N., M.P. Castles, and W.M. Sherrill, "LF to VHF Surface Ship Direction Finding Research," San Antonio, TX: Southwest Research Institute, Interim Development Report, Contract NObsr-8917, ASTIA No. AD-465 791, May 1965.

Chapter 2
OPERATING PRINCIPLES

2.1 GENERAL DIRECTION-FINDING PRINCIPLES

The primary function of a radio direction finder is to determine the direction of arrival of an incident electromagnetic wave as received at the RDF site. This chapter discusses both general radio direction-finding principles and specific direction-finding techniques used for small-aperture direction finding.

Figure 2.1 shows a representative radio direction-finding (x,y,z) spatial coordinate system with the RDF located at the $(0,0,0)$ coordinate point. The xy-plane is the azimuth or bearing angle plane. The zy- and zx-planes are vertical angle of incidence or elevation angle planes. The angles of arrival associated with the direction of arrival are the azimuth angle, ϕ, measured from the y-axis, and the elevation angle, θ, measured from the z-axis. The z-axis is usually aligned on the zenith. The y-axis is established as a reference axis of zero degrees aligned on true, magnetic, compass, or grid north as determined by operational considerations. In some situations, such as airborne applications, the azimuth plane may coincide with a reference plane of the host platform.

General direction-finding principles assume that the electromagnetic field incident on the RDF exhibits a far-field planar wave structure with linear polarization. As depicted in Figure 2.1, the radial E-field is zero. The tangential E-field, E_t, and the H-field are in space quadrature. The direction of propagation is indicated by Poynting's vector, $\mathbf{P}$.

In practice, the incident electromagnetic field is usually nonplanar with phase-front distortion, created by multipath and scattering, and depolarization, produced by nonuniform ionospheric propagation effects.

The basic RDF function is to define a radius vector $\mathbf{Q}$ that is parallel to $\mathbf{P}$ such that $\mathbf{Q}$ is in the direction of the incident electromagnetic wave. A full-capacity generic RDF should measure direction of arrival in three-dimensional space. However, many operational situations require only measurement of the directional component, ϕ, in the azimuth plane. A one-coordinate RDF system provides AOA information in only one plane—usually the azimuth plane. A two-coordinate RDF

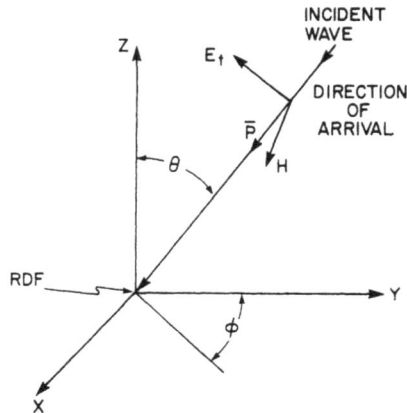

Figure 2.1 RDF spatial coordinate system.

system can provide both azimuth and elevation angle information. Most small-aperture RDF systems are one-coordinate systems providing azimuth AOA data only.

The measured angle of arrival, ϕ, is commonly called a bearing angle or line of bearing (LOB). However, a measurement of ϕ is only a bearing estimate. A strict definition [1] of the term "bearing" is as follows:

(A) The horizontal direction of one terrestrial point from another, expressed as the angle in the horizontal plane between a reference line and the horizontal projection of the line joining the two points. (B) Azimuth. A bearing is often designated as true, magnetic, compass, grid, or relative and is dependent upon the reference direction.

The true bearing to an RF source is always the great-circle bearing (GCB) because electromagnetic energy is propagated in a radial direction from an RF source. Unperturbed transmissions always follow a great-circle path, which is the shortest distance between two points measured on the earth's global surface. Therefore, in the strictest sense, the term "bearing" is actually the great-circle bearing. However, in deference to common usage, this book will use the terms "bearing" and "line of bearing" for measured bearing. Whenever true bearing or great-circle bearing is intended, they will be noted as such.

The ultimate goal of any radio direction finder is to measure the great-circle bearing of the subject RF source. However, the best performance only provides a measure of the bearing and, hence, an estimate of the great-circle bearing.

Geolocation of the RF source is of primary interest for many DF applications. The two principal operating strategies for geolocation are stand-off and homing direction finding. Stand-off geolocation from a single site requires three-dimensional measures such as range, ϕ, and θ. However, because RDFs are passive systems, range is not attainable from a single site. (Single station location can obtain range via ionospheric ray tracing.) Hence, stand-off RDF geolocation requires bearing measurements from multiple sites.

Bearing data obtained from multiple, dispersed, stand-off RDF sites (or from a single moving RDF) are used to estimate the geolocation of the RF source via triangulation techniques. Triangulation uses the bearings from two or more RDF locations to obtain an estimate of the source's geographic location. Triangulation techniques are discussed in a classical paper by Stansfield [2]. Triangulation techniques have been extensively documented and are discussed in greater detail in Chapter 7. Reference [3] contains a comprehensive discussion of emitter location.

For navigational purposes, geolocation of the RDF site may be needed. The RDF is used to obtain bearings on RF sources at known locations. The RF sources are usually cooperative in that they have been deployed especially for navigational purposes. Bearings are obtained on three or more RF source locations, and triangulation plots are made using reciprocal bearings from the known RF source locations to obtain an estimate of the RDF geolocation. This technique is widely used for air and marine navigation and is probably the most common RDF application.

Actual geolocation of an RF source may be attained by homing, which involves following a bearing (or azimuth) course directed toward the RF source. A constant match or reference is maintained on the measured bearing as the RDF is moved. The RF sources may be both cooperative and uncooperative.

The following section discusses direction-finding techniques for small-aperture radio direction finding.

2.2 SMALL-APERTURE DIRECTION-FINDING PRINCIPLES

2.2.1 Introduction

Small-aperture RDFs determine direction-of-arrival information by three measurement methods:

1. Amplitude response;
2. Phase delay;
3. Time delay.

The angles of arrival ϕ and θ are converted into voltage analogs using amplitude response, phase delay, or time delay. Basic conversion techniques are presented in the following sections.

2.2.2 Amplitude Response

Two types of amplitude response are used to obtain DF information: direct and comparative.

For direct amplitude response, the directional properties of elemental H-field loops and E-field dipoles provide response minima as they are rotated in azimuth. Bearing data are acquired at these minima response orientations.

Elemental loops and dipoles exhibit figure-eight responses as depicted in Figure 2.2 for the azimuth plane. Response maxima are broad but the minima (nulls) are sharp. The amplitude response derivative, as a function of ϕ, is greatest around the null response position. Therefore, a dynamic measure of null position provides the best bearing estimates.

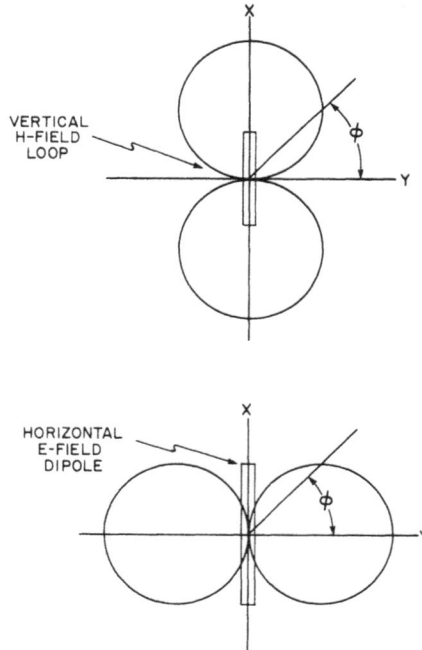

Figure 2.2 Azimuth plane response patterns.

Sequential amplitude measurements are made as the directive pattern is rotated in azimuth, and the measured bearing is indexed to the azimuth at which some predetermined amplitude-null property occurs.

The general equation for the induced voltage, V, in an electrically small, vertical loop oriented normal to the xy-plane is given by

$$V = KE[\sin\phi \cos\psi - \cos\phi \cos\theta \sin\psi] \tag{2.1}$$

where

$K = (2\pi/\lambda)(\text{area of loop, number of turns on loop});$
$E = \text{incident signal field strength (volts/meter)};$
$\lambda = \text{wavelength (meters)};$
$\phi = \text{azimuth angle between the direction of propagation and the normal to the plane of the loop (degrees)};$
$\theta = \text{elevation angle referenced to the zenith (degrees)};$
$\psi = \text{polarization tilt angle of the incident signal, the angle between the } E\text{-field vector and the vertical plane normal to the plane of the loop (degrees)}.$

Equation (2.1) shows that nonsurface wave signals ($\theta < 90°$) with horizontal polarization ($\psi > 0°$) produce bearing errors. This is a major fundamental problem of loop DF systems and is discussed in greater detail in Chapter 5.

Loop bearing measurements contain a 180° azimuth ambiguity. In some applications, this is not a problem. When ambiguity resolution is needed, the loop output is combined with the output of a vertical E-field element to obtain a cardioid response pattern with only one null and, hence, no ambiguity. Figure 2.3 depicts a cardioid response pattern. In some RDFs, the cardioid pattern is used to provide the bearing information.

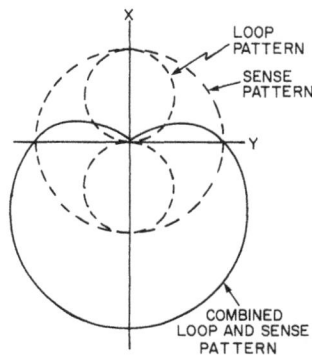

Figure 2.3 Cardioid pattern formation.

The horizontal E-field dipole is rarely used for DF purposes, primarily because most signals of interest in the HF and VHF bands are vertically polarized and of opposite polarization to the horizontal dipole.

For comparative amplitude DF, multiple antennas are used to obtain orthogonal figure-eight patterns as shown in Figure 2.4. Bearing information is obtained from the ratio of the amplitudes of the orthogonal figure-eight patterns. Using Figure 2.4, the bearing angle ϕ is given by the function as follows

$$\phi = \text{function}[G_x(\phi)/G_y(\phi)] \tag{2.2}$$

The $G_x(\phi)$ and $G_y(\phi)$ responses are converted to two voltages, V_x and V_y, which are proportional to the sine and cosine of the bearing angle ϕ. Processing the log of the ratio V_x/V_y produces the algorithm $\log(\tan\phi)$, which may be further processed to obtain the bearing angle ϕ.

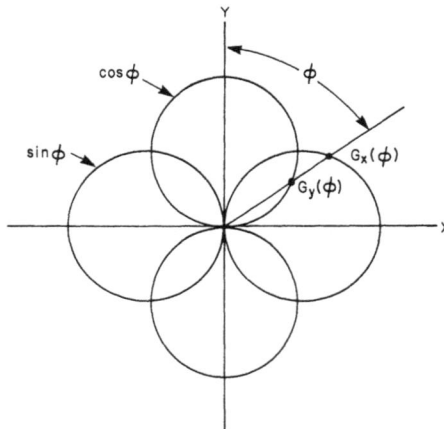

Figure 2.4 Amplitude comparison technique.

Comparative amplitude DF techniques provide instantaneous bearing information. Further, because the incident signal has been normalized by the ratio process, the effects of signal perturbations on the bearing measurements have been reduced.

Comparative DF techniques contain bearing ambiguities. Ambiguity resolution and other design considerations are discussed in Chapter 5.

2.2.3 Phase Delay

Bearing measurements using phase delay require at least two separated antennas. A plane wave, arriving at an angle other than the normal to the baseline between the two elements, arrives at one element earlier than the other. The time lag between the antennas produces a RF phase delay or a differential RF phase between the antenna outputs.

Figure 2.5 illustrates the basic phase-delay technique. An incident plane wave arrives at an incident angle ϕ at antenna 1 inducing a voltage $V_1 \exp(j\omega t)$. After propagating the distance $d \sin\phi$, the incident plane wave induces a voltage $V_2 \exp(j\omega t - \tau)$ in antenna 2, where τ is the phase delay given by

$$\tau = (2\pi d/\lambda) \sin\phi \tag{2.3}$$

Therefore, the bearing angle ϕ is encoded as a function of phase delay τ.

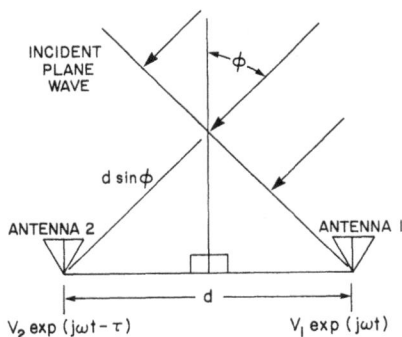

Figure 2.5 Phase-delay DF parameters.

The phase-to-amplitude DF technique is shown in Figure 2.6. The RF phase differential across the baseline is converted to an amplitude function in a 180° hybrid. This amplitude function exhibits a figure-eight response pattern for a small-aperture baseline. (This is the well-known Adcock DF technique discussed in Chapter 1.)

Figure 2.7 depicts the phase interferometry technique for measuring RF phase delay τ. The outputs of each antenna element are individually received, processed, and converted to an IF signal. The differential RF phase τ between each channel is measured by a phase comparator.

16

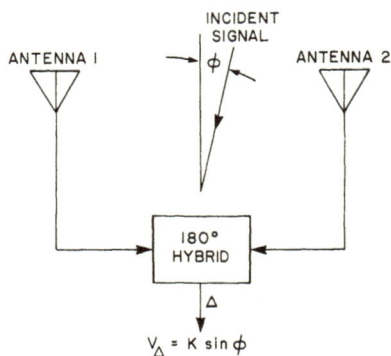

Figure 2.6 Phase delay-to-amplitude DF technique.

For both phase-delay techniques, the bearing angle ϕ is computed by using Eq. (2.3), where phase delay τ is measured and baseline distance d and wavelength λ are known.

Both phase-delay techniques can experience phase and bearing ambiguities. If $d > \lambda/2$, phase ambiguities exist; however, for small-aperture DF applications phase ambiguities are not a problem. Bearing ambiguities are resolved by using auxiliary antennas to establish known phase references. Chapter 5 presents further discussion of ambiguity resolution and other design considerations.

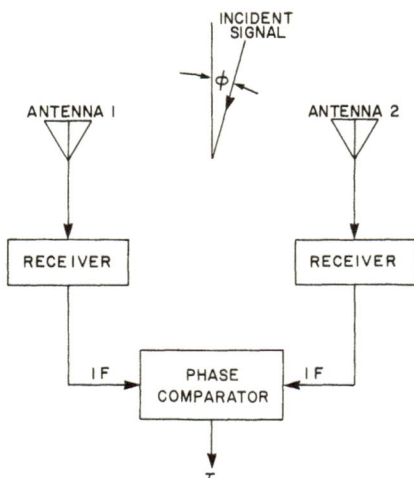

Figure 2.7 Direct phase (interferometer) DF measurement.

A variant of the phase measurement method is doppler direction finding. Doppler DF is generally considered to be a medium-aperture technique; however, its use is growing in the small-aperture area. The basic principles of doppler DF are discussed below.

The received frequency from a moving antenna experiences a doppler shift. If the antenna moves along a radial to the emitter position, the doppler shift is maximized. No doppler shift occurs if the antenna movement is tangential to the direction of propagation. These maxima and minima doppler shifts may be used to measure the bearing of the incident signal.

If the antenna rotates on a circle with radius r, the doppler frequency shift varies sinusoidally. If f_r is the rotation rate and λ is the signal wavelength, the peak doppler shift is $2\pi r f_r/\lambda$. The signal bearing is tangent to the circle of rotation at the maximum doppler point and is in the direction of antenna movement.

In practice, mechanical antenna rotation is not used, but a pseudodoppler technique is used. Pseudodoppler DF is based on sequential phase measurements from a circular array of fixed antennas sequentially switched to a common receiver. Sequential sampling provides doppler-induced frequency deviations at discrete angles. The doppler amplitudes obtained at the discrete sample points can be processed and averaged to acquire the maximum doppler shift angles and, hence, bearing measurements. The doppler DF technique is discussed further in Chapter 5.

2.2.4 Time Delay

An electromagnetic wave propagating between two separated antennas experiences a time delay. This time differential of arrival (TDOA) contains bearing angle information. In Figure 2.8, the differential time delay, Δt, between antennas 1 and 2 is given by

$$\Delta t = (d/c) \cos\phi \qquad (2.4)$$

where

d = the distance between antennas 1 and 2 (meters);
c = velocity of light;
ϕ = bearing angle (degrees).

Historically, TDOA has been used for pulsed emissions such as radar. Also, relatively long, multiple-wavelength baselines are commonly used. These factors appear to be inconsistent with small-aperture HF and VHF applications. However, the advent of HF and VHF transmissions with pulsed properties and the improved

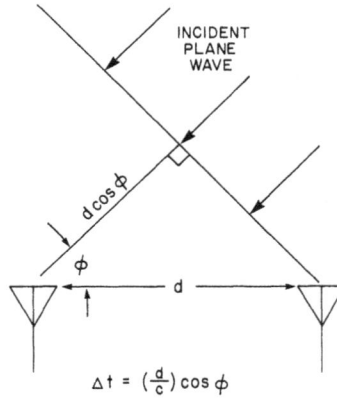

Figure 2.8 Differential time-of-arrival parameters.

capabilities for precision time-interval measurements are increasing the potential for TDOA application to small-aperture HF and VHF direction finding. Chapter 5 provides an expanded discussion of TDOA for small-aperture use.

REFERENCES

1.   Jay, Frank, ed., "IEEE Standard Dictionary of Electrical and Electronic Terms," ANSI/IEEE Standard 100-1988, New York: Institute of Electrical and Electronic Engineers, Inc., 1988.
2.   Stansfield, R.G., "Statistical Theory of DF Fixing," *Proc. IEEE,* Vol. 94, Part 3A, No. 15, 1947, pp. 762–770.
3.   Wiley, R.G., *Electronic Intelligence: The Interception of Radar Signals,* Norwood, MA: Artech House, 1985, Chapter 5.

Chapter 3
PERFORMANCE DEFINITION

3.1 PERFORMANCE CATEGORIES

Planning for radio direction-finder design, development, and application should include a thorough definition of required technical and operational performance parameters. The structure of a parameter set is a function of the specific application, which may be highly specialized. However, formulation is possible of a parameter set that includes a large majority of the factors that should be considered for any given application. This generic parameter set treats both technical and operational requirements and operating capabilities. Both quantitative and qualitative parameter definitions are pertinent; however, quantitative definitions are desired when feasible.

Comprehensive parameter characterization is especially vital if the contemplated RDF is to be used against uncooperative RF sources. If the RF source is cooperative, RDF parameters are generally standardized and application oriented. The RDF user or planner is presented with a parameter set that is already well defined. The RDF selection process then involves parameter assessment and comparative evaluation rather than parameter definition. For example, aircraft RDFs are usually used on an established net of nondirectional radiobeacons (NDBs) with stylized, controlled, uniform transmitted signal characteristics. Further, aircraft RDF manufacturers adhere to minimum technical and operational performance standards formulated by organizations such as the Radio Technical Commission for Aeronautics (RTCA) [1]. Hence, from the user's viewpoint, the selection of an aircraft RDF may be based solely on specific performance features or economic reasons.

The following discussion emphasizes parameter categories for RDFs used against uncooperative emitters. Radio direction finder performance may be categorized as follows

1. Electrical performance;
2. Physical characteristics;

3. Environmental and service conditions;
4. Interface functions.

The *electrical performance* category specifies performance in the frequency, signal, space, and time domains. This category generally establishes the basic DF technique required and significantly influences definition of the other three categories.

The electrical performance category often includes internal and external power requirements, self-test and built-in test capabilities, and fault isolation.

The *physical characteristics* category includes weight, volume, form factor, dimensions, and special requirements, such as subsystem modularization and collapsible antennas. If antenna mast installation is planned, the physical characteristics of the antenna subsystem must be carefully defined because mast installation can be a major cost driver.

The *environment and service conditions* category includes factors such as temperature, altitude, humidity, shock, vibration, and wind loading and icing (on the antenna subsystem). Also, this category may include the ". . . *ility*" requirements such as the following:

- Reliability;
- Maintainability;
- Transportability;
- Testability;
- Deployability.

This category also includes specification of the siting effects of the host platform or local environs as well as the land area needed and counterpoise requirements (for ground-based systems).

The *interface functions* category addresses interactions with ancillary subsystems, such as ionosounders, and control–data functions. Generally, the control-data interface requirements are defined by specifications such as RS-232, IEEE-488, ARINC 429 (for commercial aircraft RDFs), and MIL-STD-1553B (for military RDFs).

3.2   MAJOR PARAMETERS

Major parameters drive DF technique selection, establish design and development objectives, and significantly influence subsystem specification. These key parameters fall in the electrical performance category and are associated with the frequency, signal, space, and time performance domains.

3.2.1   Frequency Domain

Overall, frequency-domain parameters have the greatest impact on RDF design, implementation, and operational effectiveness. Pertinent frequency-domain parameters are the frequency region of coverage, total RF band coverage (extent of coverage), and the instantaneous RF bandwidth required.

The frequency region of operation plays a major role in determining the signal-domain characteristics that may be encountered and, hence, the propagation error sources that an RDF will experience. The total RF range of coverage may necessitate broadband operation. Sensitivity and bearing accuracy can be adversely affected by wide RF coverage (broadband operation), especially if an untuned antenna subsystem is used. If a tuned antenna subsystem is used to mitigate sensitivity degradation, signal and bearing acquisition times may increase due to the increased time required for bandswitching and retuning as the operating frequency is changed.

For DF techniques using disposed antenna elements (i.e., baseline techniques), total RF coverage in the 10:1 range may require the use of multiple antenna bays.

Wideband transmissions produced by modulations such as spread spectrum, frequency hopping, and burst modes require a broadband instantaneous RF bandwidth, which, in turn, increases antenna and RF circuitry complexity. Specifically, broadband antenna elements and wideband linear amplitude and phase RF circuitry may be required.

Table 3.1 shows representative small-aperture RDF applications. It illustrates some of the major interactions between frequency-domain characteristics and DF techniques, antenna types, RF source class, and application areas.

Table 3.1
Representative Small-Aperture Direction Finder Applications

RF Source	Application	Frequency (MHz)	DF Technique	Antenna
Cooperative	Marine HF	0.2 to 18	Amplitude	Tuned crossed loops
	Marine VHF	156 to 162	Phase to amplitude	Untuned dual-baseline array
	Aircraft HF	0.19 to 1.75	Amplitude	Tuned crossed loops
Uncooperative	Surveillance HF–VHF	20 to 200	Phase	Untuned dual-baseline two-bay array
	Surveillance MF–HF–VHF	0.5 to 100	Amplitude	Single tuned loops [five required]

3.2.2 Signal Domain

Signal-domain characteristics that affect DF technique selection the most are modulation, transmission duration, and plane-wave distortion imposed on the signal by external factors.

For DF applications using cooperative RF sources, signal-domain characteristics are controlled to enhance DF performance. However, for uncooperative RF sources, deleterious signal-domain effects may control DF technique selection and hardware implementation.

Modulation compatibility is a major requirement for effective RDF operation against uncooperative emitters. Generally, RDFs use DF techniques that are compatible with a wide variety of modulations including both the conventional CW, AM, FM, SSB, and narrowband pulse modulations as well as the noise-like modulations such as spread spectrum and burst transmissions. However, notable incompatibilities do exist. For example, amplitude DF techniques exhibit inherent limitations when used against carrier-less emissions such as SSB. Also, amplitude DF techniques using noninstantaneous bearing acquisition methods (scanned antennas) manifest degraded performance on SSB, short-duration, and burst transmissions. Doppler and pseudodoppler DF techniques have very limited capability on short-duration transmissions, noise-like spread-spectrum emissions, and SSB modulations.

RDFs operate best on signals with plane wavefronts and fixed polarization. In practice, signals are not ideal—wavefronts are nonplanar and depolarization occurs. Small-aperture direction finders operate primarily in the MF, HF, and VHF bands where significant signal wavefront distortion and abnormal polarization can occur. The operating frequency band, its extent, and the maximum range to the RF source have a major impact on introducing signal-domain imperfections and, hence, degraded performance. Signal-domain imperfections must be carefully defined and included in RDF planning and application; these imperfections are discussed in greater detail in Chapter 5.

3.2.3 Space Domain

The required field of view (FOV) for an RDF specifies performance in the space domain. FOV parameters are usually specified for both the azimuth and elevation planes. The FOV may be specified as either sequential or instantaneous. A sequential FOV is covered by a scanned antenna. An instantaneous FOV is covered by an antenna subsystem that responds instantaneously to signals arriving from any point within the specified FOV.

A full and instantaneous FOV necessitates continuous 360° hemispherical coverage in both the azimuth and elevation planes for airborne applications. If the RDF is surface-based, a full and continuous FOV requires 360° azimuth and $\pm 90°$

semihemispherical elevation coverages. Full and continuous FOV coverage in any plane requires the use of baseline instantaneous DF techniques and precludes the use of single-element DF methods and antenna scanning. Full and continuous FOV coverage in both the azimuth and elevation planes requires, at least, dual baselines and nonscanned antenna methods.

The use of cooperative RF sources places fewer demands on space-domain requirements. FOV requirements may be tailored to match the known or likely locations of the RF source. For example, an aircraft emergency locator transmitter (ELT) may be classified as a cooperative source. In use, its exact location is certainly not immediately known; however, we may assume it is on the earth's surface. Therefore, an RDF for ELT sources needs only a relatively small elevation FOV and 360° azimuth FOV.

For RDFs used against uncooperative sources, a requirement is arising for full and instantaneous FOV coverage capability due to the increase in short-duration transmissions and spread-spectrum modulations.

3.2.4 Time Domain

The time required to perform signal intercept and acquire bearing information establishes time-domain requirements. An operationally unacceptable situation exists if the total required time for intercept and bearing acquisition exceeds the nominal transmission time of the signal of interest. In general, time-domain factors are important only if the RF source is uncooperative. Cooperative RF sources relieve time stress on the RDF by providing either continuous or high-duty-cycle transmissions.

Signal intercept time and bearing acquisition are often specified separately. Signal intercept time may include not only the time to tune and detect the signal of interest, but also the time required for signal identification and DF net synchronization. Bearing acquisition follows signal intercept. The bearing acquisition time may include the time required to accomplish DF antenna and receiver tuning and perform bearing error-reduction processing.

RDF time-domain parameters must be carefully delineated and defined due to the many diverse factors involved and the adverse operational implications of erroneous specification. All steps in the signal intercept and bearing acquisition process must be included, and a comprehensive time-budget must be formulated. Statistical time-domain parameters are very likely to be necessary.

REFERENCES

1. "Minimum Operational Performance Standards for Automatic Direction Finding (ADF) Equipment," Document No. RTCA/DO-179, Special Committee 146, Radio Technical Commission for Aeronautics (RTCA), May 1982.

Chapter 4
DIRECTION-FINDING ERROR SOURCES

4.1 INTRODUCTION

Small-aperture direction finders are susceptible to a wide variety of direction-of-arrival error sources, which may be divided into four major categories:

1. Propagation-induced errors;
2. Environmental errors;
3. Instrumental errors;
4. Observational errors.

As a signal travels from an RF source to a DF system, direction-of-arrival errors accrue in the sequence listed above. The errors are cumulative because they originate from independent causes that involve statistical definitions. The error statistics for each error category are statistically combined to arrive at an overall error definition.

After transmission, a signal traverses the propagation medium, such as the ionosphere, which introduces direction-of-arrival deviations, time dispersion effects, and signal amplitude degradation. Next, as the signal approaches the DF site, environmental errors occur when the signal encounters natural and man-made scattering and reradiation objects such as power lines, towers, ship masts, dense vegetation, *et cetera,* that deviate the signal from a linear propagation path. At the DF system, the signal may experience instrumental errors caused by equipment imperfections such as antenna unbalance, antenna pattern distortion, and amplitude and phase unbalance in multiple-channel receivers. Finally, the direction-of-arrival information may be degraded by observational errors introduced at CRT displays or aural-null interfaces.

The four error-producing categories will be discussed in the following sections with emphasis on error mechanisms and their relative effects.

4.2 PROPAGATION-INDUCED ERRORS

4.2.1 Propagation Mechanisms

Propagation-induced errors are those created by the intervening media as the signal of interest propagates between the RF source and the DF site. Electromagnetic waves propagate from a RF source to a DF site in four major ways: ground, iono-spheric (sky wave), tropospheric, and scatter propagation. Figure 4.1 depicts the various wave components excluding the surface and scatter modes. The ground wave travels directly from the RF source to the DF site, whereas the sky wave reaches the DF site via refraction from the ionosphere. (The term "sky wave" will be used rather than "ionospheric wave." The use of the term "ionospheric wave" is deprecated because it is now intended to connote wave characteristics in iono-spheric plasma [1].)

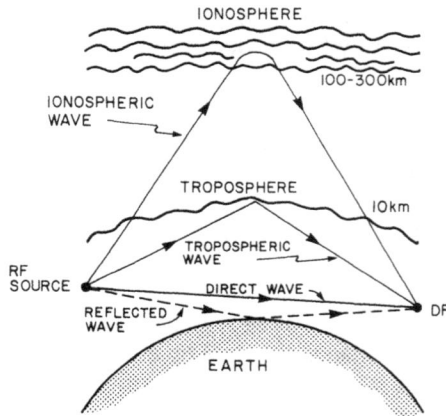

Figure 4.1 Propagation mechanisms.

The tropospheric wave generally propagates along the great-circle plane via refraction from tropospheric irregularities. Both ionospheric and tropospheric scat-ter can occur and can exhibit considerable deviation from the great-circle path. All scatter modes manifest rapid, flutter fading.

Ground Wave

The ground wave is composed of one or more of the following wave components: direct wave, ground-reflected wave, and the surface wave. The direct wave, which is unaffected by the earth's surface, is primarily an air-to-air or air-to-ground mode as is the ground-reflected wave. At the DF system, the direct and ground-reflected waves combine in phase and out of phase to produce amplitude fading, which is particularly significant when the source or DF platforms are moving. Both the direct wave and the ground-reflected wave obey the inverse law of radiation with the transmission loss varying inversely with path distance.

When both the RF source and the DF site are on or close to the earth's surface, the direct and ground-reflected waves are near RF phase opposition and tend to cancel, leaving only the surface wave transmission mode (assuming no sky wave or tropospheric wave components are present).

The surface wave is guided along the earth's surface and is affected by the conductivity and dielectric constant of the earth along the path. Surface wave propagation has been extensively treated in the literature [2–4]. Prediction methods are well formulated [5–9], including those for forested as well as urban areas [10–12].

Below about 30 MHz, the surface wave is more significant than the direct and reflected waves except when dealing with air-to-air and air-to-ground transmissions. The effects of frequency, earth conductivity, and dielectric constant on surface wave propagation are very significant. The rate of surface wave attenuation increases as frequency increases and decreases as conductivity increases. The effective range of the surface wave is much greater over seawater than over land, because the conductivity of seawater is some 1000 times greater than typical soil. The greater surface wave range over seawater contributes to the effectiveness of marine RDFs operating in the MF band.

Above 30 MHz, the direct and reflected waves are the major components of the ground wave. For extended over-the-horizon transmissions, the troposcatter mode dominates.

In dense forests, a propagation mode has been found to exist along the tree tops at the forest-air interface. This mode, called the "lateral wave," occurs from about 1 to 100 MHz.

Clearly, since small-aperture DF usage extends from VLF to above VHF, the DF practitioner may have to deal with a wide variety of ground wave components. Fortunately, most components are well behaved from a direction-of-arrival viewpoint except for the lateral wave mode that may experience significant deviation from the great-circle path.

Sky Wave

Sky waves can exhibit major direction-of-arrival errors especially when received by small-aperture direction finders. The cause and magnitude of the errors are highly variable and are significantly affected by space-, time-, and frequency-domain parameters. The purpose of this section is to summarize the essential elements of sky wave propagation with emphasis on those properties that have a major impact on DF errors.

The principles and practices of sky wave propagation are extensively documented. DF practitioners operating at frequencies below about 30 MHz are encouraged to acquire in-depth knowledge of ionospheric propagation and especially the effects on DF direction-of-arrival errors. Reference [13] provides a detailed treatment of ionospheric radio propagation and the mechanisms that create DF errors.

The ionosphere is defined [1] as follows:

(1) That part of the earth's outer atmosphere where ions and free electrons are normally present in quantities sufficient to affect propagation of radio waves. (2) That part of a planetary atmosphere where ions and free electrons are present in quantities sufficient to affect the propagation of radio waves.

In the ionosphere, the constituent gases of the atmosphere are ionized by radiation from outer space—especially ultraviolet light from the sun. The major portion of the ionosphere exists from an altitude of about 40 to 300 km, and it consists of three major layers called the D-, E-, and F-layers. Actually, the ionosphere is a continuum of ionized gases, and no true layering exists. The "layers" are ionization density maxima that vary hourly, daily, annually, and cyclically over multiyear time periods such as the 11-year sunspot cycle.

Table 4.1 presents pertinent characteristics of the layers. During the day, the F-layer separates into the F1- and F2-layers. The "maximum" refraction frequencies listed are approximate, increasing and decreasing in a manner proportional to sunspot number. The D-layer is the major absorption layer for the MF and HF bands; hence, the D-layer decreases the intensity of signals refracted from higher layers. The E- and F1-layers simply attenuate signals passing through to the higher layers. At night, when the D- and E-layers do not exist, the strength of sky wave signals increases dramatically.

Table 4.1
Characteristics of the Ionospheric Layers

Layer	Nominal Height (km)	Absorbs	Attenuates	Refracts	Diurnal Life
D	40 to 90	MF/HF	—	VLF–LF (to 500 kHz)	Day
E	90 to 130	—	MF–HF	MF–HF (to 20 MHz)	Day
F1	175 to 250	—	MF–HF	MF–HF (to 30 MHz)	Day
F2	250 to 400	—	—	—	Day
F	270 to 400	—	—	MF–HF (to 30 MHz)	Night

In addition to the five major ionospheric layers, several transitory, irregular layers exist, with the sporadic-E (Es) being the most important. The Es-layer consists of relatively dense patches of ionization at heights between 95 to 120 km. The occurrence and character of Es is highly variable and relatively unpredictable. Due to a high critical frequency created by dense ionization, Es provides for relatively long-distance communications in the VHF band at frequencies up to 80 MHz. Unfortunately, sporadic-E can be a major source of DF errors in the HF band under certain conditions.

The ionospheric layers are continuums of ionized gases covering a region of considerable extent; however, for modeling and prediction purposes, definition of a virtual height for each refractive layer is convenient. Virtual heights are determined empirically by use of ionosoundes (ionosounders). Ionosoundes use either pulsed or swept-frequency RF to measure the amplitude and time delay of refracted signals as a function of frequency. The ionosounde directs energy toward the zenith and measures the time required for a round-trip path. This time is then converted to virtual height assuming the velocity of propagation is the velocity of light in a vacuum over the entire path.

In Figure 4.2, the virtual height is that height indicated by ionosounde measurements. A signal incident on the ionospheric layer at an angle θ_0 is refracted along the path B or B', which has approximately the same time delay as a signal reflected from a perfect conducting layer with an edge at the height A or A'. Because ionospheric group velocity is less than the group velocity in a vacuum, the time delays of paths A or A' and B or B' are approximately equal, and A (A') is the virtual height.

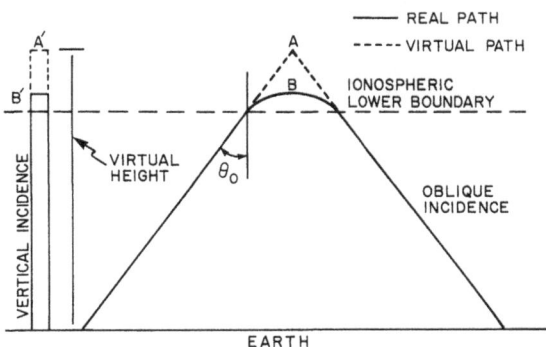

Figure 4.2 Depiction of vertical and oblique incidence sky waves.

For a vertically incident signal ($\theta_0 = 0°$), as frequency is increased, a frequency is reached at which no signal will be refracted by the ionospheric layer, but the signal will penetrate the layer and not return to earth. This frequency is called the critical frequency, *fc*. Note that *fc* is not the highest frequency that can be returned from the layer. The highest frequency is a function of the oblique angle of incidence θ_0, which, in turn, is dependent on the distance between the transmitter and receiver. For any given layer, the highest frequency that can be refracted to the earth for a given distance of transmission is the maximum usable frequency (MUF). The MUF is a function of the critical frequency and angle of incidence θ_0 by the expression

$$\text{MUF} = (fc)(\sec\theta_0)$$

The skip distance is that distance between two points on the earth's surface for which a given frequency is the MUF. Within the skip distance, no ionospherically refracted signals are received; however, ground waves may be received. The region between the ground wave coverage extent and the edge of the refracted signal zone is called the skip zone as illustrated in Figure 4.3.

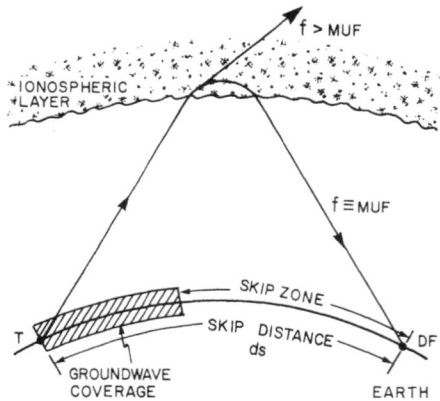

Figure 4.3 Skip-zone characteristics.

The skip distance is a complex function of the ionosphere ionization state, angle of incidence θ_0, and the operating frequency. The skip distance increases as frequency increases. At frequencies below about 5 MHz and F-layer refraction, the skip distance is negligible and NVIS is feasible. In the 5- to 7-MHz region, the skip distance becomes appreciable, e.g., several hundred kilometers at night. Above about 10 MHz, the skip distance can extend to several thousand kilometers. For the DF practitioner, the skip zone is very important and is discussed in Section 4.2.2.

MF and HF communicators attempt to operate links at the optimal working frequency (OWF or FOT from the French initials). The OWF is set as some frequency laying between about 0.5 to 0.85 of the monthly median MUF to allow for day-to-day variations in the critical frequency. The use of the term "optimal" is misleading in that the OWF is not necessarily the frequency providing maximum signal strength or minimum time dispersion; however, the OWF is the frequency that tends to produce reliable communication on a statistical basis. The MUF may exhibit daily variations about the monthly average of up to 15%; use of the OWF provides the statistical margin needed for reliable, consistent communication.

As the operating frequency is reduced from the MUF, ionospheric absorption increases and the received signal strength decreases. The lowest usable frequency (LUF) is the minimum frequency below which signal strengths are unacceptable. Unlike the MUF, which is a function of ionospheric conditions and distance only, the LUF is a function of ionospheric absorption, transmitter effective radiated power, and the required signal strength (and signal-to-noise ratio) for the subject communication link. Hence, total path loss and the received noise level are major LUF parameters. The LUF is maximum around local noon.

Figure 4.4 shows sample diurnal plots of MUF and LUF for an exemplary transmission path. At night, 2 MHz is a reasonable OWF; however, during the day, 2 MHz lies below the LUF, and an OWF of about 6 MHz is a good choice.

The magnetoionic propagation mode of the F-layer is especially important to DF users due to associated error-producing effects. In the presence of the earth's magnetic field near the critical frequency, the ionosphere decomposes an incident

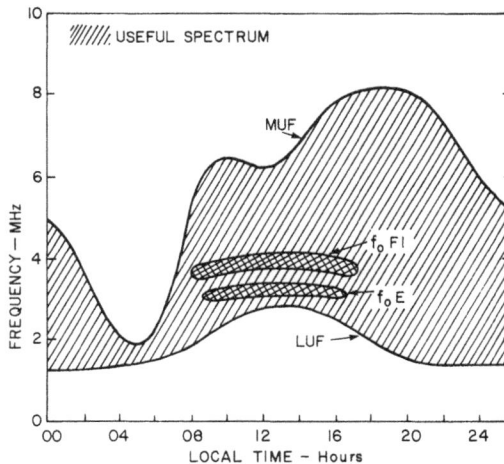

Figure 4.4 MUF and LUF plots.

plane-polarized wave into ordinary and extraordinary waves (magnetoionic components) and deviates their propagation paths from the great-circle plane, creating the potential for significant DF bearing errors. Further, the ordinary and extraordinary waves are essentially circularly polarized and experience different phase velocities. When the ordinary and extraordinary waves emerge from the refracting layer, they combine to form a resultant elliptically polarized plane wave with rotating polarization. This polarization rotation, called Faraday rotation, is a source of bearing errors in DF systems using antennas that are polarization sensitive, for example, a loop antenna. Faraday rotation errors are considered instrumental errors.

The existence of ionospheric irregularities can also have a major impact on DF performance. Two major types of irregularities are ionospheric layer tilt and traveling ionospheric disturbances (TIDs). Both types introduce single-mode direction-of-arrival errors. In addition, they also create multiple refractions from wave-like or corrugated layers that result in multimode propagation. Multimode propagation then produces multipath DOA errors.

Ionospheric tilts are horizontal gradients in the layer electron density that create nonspherical layer stratification. Primarily an F-layer effect, tilts normally occur near the "gray line," i.e., the day-night transition near the terminator. Consequently, ionospheric tilts occur over a wide area (thousands of kilometers).

Traveling ionospheric disturbances are the result of acoustic-gravity wave propagation in the atmosphere that couples into the ionospheric structure. Large-, medium-, and small-scale TIDs have been delineated.

Ionospheric tilts and TIDs are not anomalies; their spatial and temporal characteristics are comparable to the "normal" features of the ionosphere. For example, Table 4.2 presents spatial and temporal decorrelation parameters for the major elements of the ionosphere.

Table 4.2
Ionospheric Spatial and Temporal Decorrelation Parameters

Layer	Spatial Decorrelation	Temporal Decorrelation
Normal layers (E, F1, and F2)	Thousands of kilometers	Hour or longer
Sporadic-E	Size dependent; approximately hundreds of kilometers	Hours to minutes; dependent on size of sporadic region
Ionospheric tilt	Hundreds to thousands of kilometers	0.5 hours or longer
Large TID	Thousands of kilometers	0.5 hours to several hours
Medium TID	50 to 100 kilometers	Approximately 5 minutes

Sudden ionospheric disturbances are the result of solar flares emitting strong bursts of ionizing radiation into the ionosphere. The major effect is a sudden, abnormal increase in D-layer ionization, which, in turn, greatly increases D-layer absorption leading to sky wave transmission "black-out"—especially on frequencies greater than about 1 MHz. (LF and VLF transmissions using the D-layer for earth-ionosphere moding may actually experience enhanced signal strength when sudden ionospheric disturbances occur.) Sudden ionospheric disturbances (SIDs) occur over the entire earth; however, ionospheric disturbances occur that are unique to the higher latitudes. They are polar cap absorption (PCA) and auroral absorption events, which result in considerable transmission loss. Direction-of-arrival deviations have not been systematically investigated for these events.

Tropospheric Reflection and Refraction

Rapid height variations of the effective dielectric constant in the troposphere create reflected tropospheric propagation. Steep gradients or abrupt changes in atmospheric humidity, temperature, and density, as influenced by the weather, are the major factors in creating rapid dielectric constant variations. Reflected tropospheric propagation is often called troposcatter even though it is a reflection process.

Tropospheric refraction is a gradual bending of an electromagnetic wave as it passes through an atmosphere with a changing index of refraction. Signals can be propagated via atmospheric ducting where the duct extends from the level of a local minimum in the index of refraction down to the level where the minimum value occurs again or down to the surface of the earth if a minimum level is not reencountered. Reference [14] presents a detailed examination of atmospheric ducting. Tropospheric reflection and refraction occur at frequencies above about 50 MHz.

Scatter

Scatter propagation is a process in which an electromagnetic signal is dispersed in direction due to scattering sources in the troposphere and ionosphere or on the earth's surface. Scatter propagation involves considerable angular dispersion of the incident wave, i.e., the direction of arrival is disordered and diffused with no significant specular component.

Scattering sources are inhomogeneities in the ionosphere or troposphere, the earth's surface, and relatively large man-made objects such as aircraft.

Ionospheric scattering is most prevalent from about 25 to 100 MHz, and is mainly the result of irregularities in the electron density distribution in the ionosphere. The major sources of irregularities are as follows [13]:

1. Turbulent D-layer mixing;
2. Spread-F;

3. Sporadic-E;
4. Meteoric ionized trails.

These sources produce the following transmission effects:

1. The D-layer produces scattering over the 30- to 70-MHz range with path lengths of 1000 to 2000 km.
2. Spread-F affords scattering up to about 50 MHz to distances greater than 4000 km.
3. Sporadic-E functions to about 80 MHz up to distances of 2000 km.
4. Meteor scatter affords 40- to 80-MHz propagation over path lengths up to 2000 km.

VHF scatter can be a major problem for DF systems; however, it is easily recognized by its rapid, flutter fading characteristics.

Tropospheric scatter produces weak but reliable transmissions from about 40 to 400 MHz over a maximum range of about 1000 km. Efficient troposcatter requires that the transmitting and receiving sites illuminate a common atmospheric scatter volume about 10 km above the earth. High-gain directional antennas are required at both the transmitting and receiving sites. Consequently, the fact that small-aperture DF systems can be consistently reliable when used on troposcatter transmissions is highly problematic.

Note that the earth's surface is an effective scatter source at MF and HF. A transmission operating over a single-hop, F-layer path near the MUF can produce considerable ground scatter. Direction-of-arrival characteristics have not been systematically investigated.

4.2.2 Direction-of-Arrival Errors

The ionosphere is a blessing to the MF–HF communicator because it affords reasonably reliable long-distance communication. However, ionospheric effects are more a bane than a blessing to the DF practitioner—especially the small-aperture DF user. The MF–HF communicator can often use real-time ionosounde data and a known transmitter location to "optimize" the transmission link. If ionosounde data are unavailable, computer-based prediction models may be used, with a known transmitter location, to establish reasonably reliable transmission links. Unfortunately, the small-aperture DF user must deal with ionospheric propagation under decidedly less than optimal conditions, and will probably experience significant ionospherically induced direction-of-arrival errors that are possibly irreducible.

This section discusses the direction-of-arrival error sources in small-aperture DF systems as a function of frequency band and, in some cases, distance from the RF source.

Very Low Frequency (3 to 30 kHz) and Low Frequency (30 to 300 kHz)

This frequency range includes the very important marine and aircraft ADF band operating from 190 kHz up. Otherwise, the VLF–LF bands offer little application for small-aperture DF. One notable exception is the use of crossed-loop–vertical monopole DF systems to locate whistlers generated by lightning impulses (sferics) that are propagated via the earth-ionosphere waveguide.

Ground wave propagation errors are insignificant compared to instrumental errors; however, a combination of ground and ionospheric (waveguide) modes can produce tens of degrees of azimuthal angle deviation [15]. Measurement techniques are subject to instrumental errors created by sky wave depolarization (Faraday rotation); therefore, the dominant VLF–LF DF error source is considered to be instrumental rather than propagation induced. Multipath conditions are created by multimode propagation paths from the earth-ionosphere waveguide duct exit point to the DF site. Multiple paths distort the incident electric and magnetic fields and preclude accurate DOA measurements by small-aperture DF systems. Section 4.2.3 discusses multipath errors in greater detail.

VLF–LF surface waves can experience wavefront distortion created by objects such as overhead power lines, towers, and extended discontinuities in ground dielectric constant and conductivity. VLF–LF direction-of-arrival deviations caused by propagation across mountain ranges and coastlines have been measured [15]. A high dependence on path characteristics and transmission geometry has been observed.

Overall VLF–LF direction-of-arrival error budgets are dominated by inherent instrumental errors created by Faraday polarization rotation [16]. An attempt to use extended baseline phase interferometer measurements with greatly reduced polarization error met with only limited success [15].

Medium Frequency (0.3 to 3 MHz)

The MF band (0.3 to 3 MHz) is populated by well-disciplined, long-established services such as the marine and aircraft beacon system (300 to 550 kHz), AM broadcast (530 to 1605 kHz), fixed and mobile communication (1.605 to 3.0 MHz), and distress and emergency calling (2.17 to 2.194 MHz). The MF band is often divided into three major service categories: beacon (300 to 550 kHz); broadcast (550 to 1605 kHz); and communication (1.605 to 3.0 MHz). (The communication service is primarily marine transmissions.) From a DF viewpoint, all operations are generally accomplished via cooperative transmissions of extended (or continuous) transmission duration. Further, the most used propagation mode is via the surface wave with vertical polarization. During the day, the sky wave is highly attenuated due to absorption in the D-layer. At night, the D-layer absorption decreases, and the sky wave becomes significant. Sky wave and surface wave interference and, hence, fading can occur. Sky wave and surface wave interference tends to be most severe at those distances from the RF source where the sky wave and surface waves are of comparable strength. Figures 4.5 and 4.6 [17] depict sky wave and surface wave regions as a function of distance and frequency for the specified conditions. The shift from daytime surface wave to nighttime sky wave conditions in the MF band is dramatic and indicates that DF conditions can change rapidly.

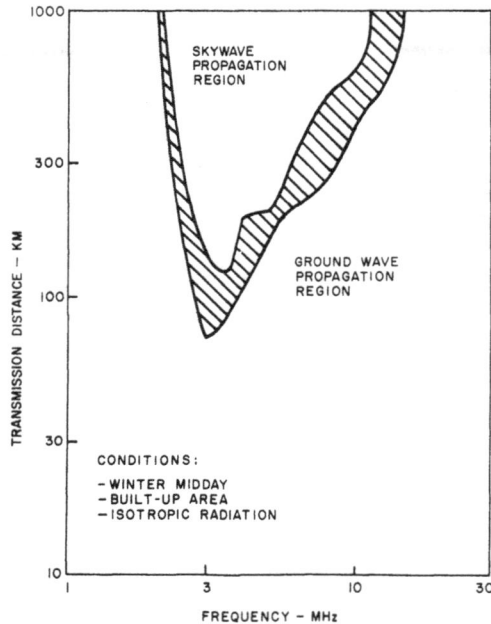

Figure 4.5 Ground and sky wave propagation ranges for winter day conditions [17].

Figures 4.5 and 4.6 were computed based on propagation through a built-up area, such as an urban area, which attenuates the surface wave more than an area that is not built up, such as water. For example, over water, the surface wave could triple in strength. The U.S. Coast Guard states that the effective range of their marine band NDB system is approximately 300 km at nighttime. An accuracy of 5° rms is cited [18].

Statistically, MF sky wave amplitudes exhibit a Rician distribution, which is a combination of a Rayleigh distribution and a specular component. Fade rates are typically of the order of 0.01 fades per second, which are considered "long" fades.

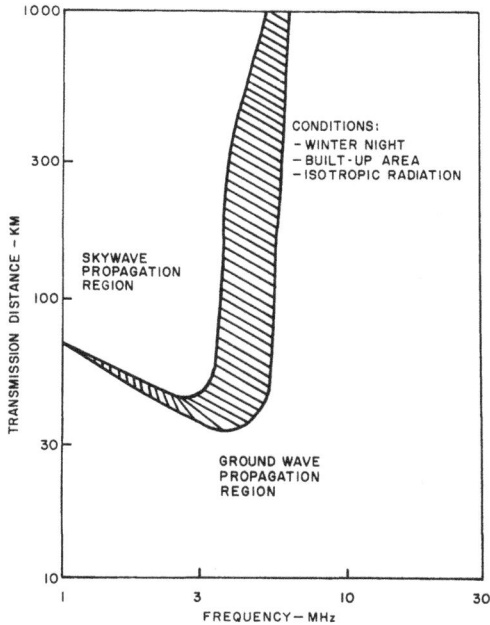

Figure 4.6. Ground and sky wave propagation ranges for winter night conditions [17].

H-field elements, such as simple loop and crossed-loop DF antennas, are preferred for MF DF use due to the need for space-efficient antennas even though *H*-field elements are susceptible to Faraday rotation polarization errors.

Users of the MF band, including DF practitioners, must contend with a very high ambient RF noise level—both natural and man-made. Representative ambient noise figures are shown in Figure 4.7. In the MF band, man-made noise levels dominate, and noise figures exceed 60 dB. To counter the high noise levels, MF services use narrowband modulations; therefore, DF systems operating on an uncooperative MF emitter should operate in a narrowband mode and, hence, pay careful attention to local oscillator frequency stability.

The propagation direction of MF surface waves can be diverted by extended discontinuities in surface conductivity or dielectric constant. For example, the conductivity of seawater is approximately 1000 times greater than the conductivity of average soil, hence, the velocity of an electromagnetic (EM) wave over sea is greater than the velocity over land. Consequently, when an EM wave traverses a coastline at an oblique angle, the wave's direction of arrival is diverted as illustrated in Figure 4.8. In Figure 4.8, wave deviation makes the boat with the DF appear closer to land than it actually is. "Shore effect" errors as large as 10° have been cited [19]. Errors are largest when the coastline elevation is high relative to the shore.

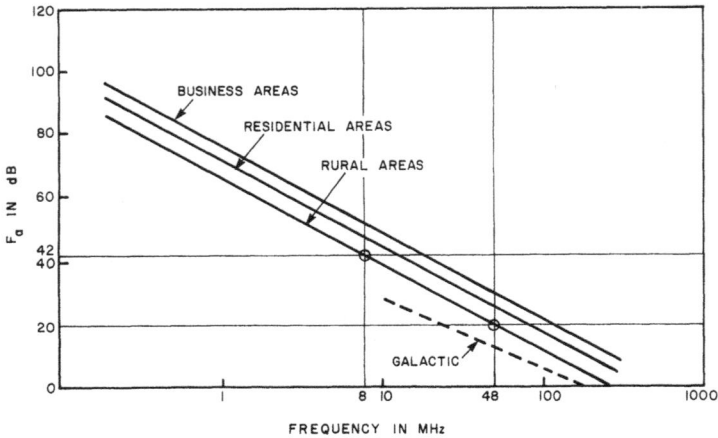

Figure 4.7 Representative noise figures (F_a).

The potential for very large DOA errors exists in the MF band. However, due to the regimentation of the services involved and the relatively high level of competence of the users, the actual operational accuracy of MF DF use is excellent.

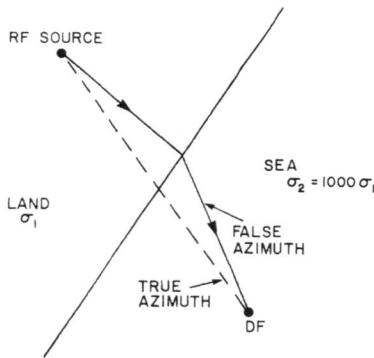

Figure 4.8. Wave deviation by land-sea interface.

High Frequency (3 to 30 MHz)

In the HF band, signal characteristics are a major function of the range between the RF source and the DF site. Similarly, direction-of-arrival errors are a major function of range. Further, DF performance, as a function of range, is also highly dependent on the DF technique used. The interaction between range and DF technique is an essential element in defining HF DF accuracy.

HF signal characteristics may be defined as a function of range to illustrate direction-of-arrival error sources. Table 4.3 defines HF signal zones and delineates the major error sources and primary effects on the DF system. In Table 4.3, the zones are listed in the order encountered as the distance between the RF source and DF site increases.

Table 4.3
HF Direction-of-Arrival Error Sources by Signal Zone

Signal Zone	Major Error Source(s)	Primary Effect
Surface wave	1. Extended discontinuities in surface electrical properties	1. DOA error
Combined ground and sky waves	1. Wave interference	1. Rapid, deep fading
	2. Magnetoionic wave splits	2. DOA offset error
	3. Ionospheric tilts	3. DOA lateral deviation
	4. Faraday rotation	4a. Instrumental polarization error
		4b. Polarization fading
Skip	1. Scatter	1a. Weak signals
		1b. Major DOA error
Sky wave	1. Multimode wave interference (multipath)	1. Fading and DOA error
	2. Magnetoionic wave split	2. DOA offset error
	3. Ionospheric tilts	3. DOA lateral deviation
	4. Faraday rotation	4. Instrumental polarization error
	5. TIDs	5. DOA offset error
	6. Sporadic-E	6. DOA offset error
	7. SIDs	7. Severe signal absorption

The HF surface wave propagation direction can be altered by extended discontinuities in surface electrical properties associated with both natural and man-made features and objects. At HF, surface irregularities and obstacles are often the same order of magnitude as the wavelength. Hence, propagation path scattering and reradiation can be localized, and relatively short discontinuities, such as singular mountains and tree lines, can have significant effects on HF signals.

Propagation deviation can occur as HF signals propagate across built-up areas such as cities and towns [11, 12, 17] and heavy vegetation [20, 21]. Surface wave path deviation is essentially random as a function of frequency, DOA angle, and distance to the source. Consequently, no deterministic method exists for predicting errors in surface wave bearing angles. The DF user should be aware that the reception of a strong, steady HF surface wave does not indicate the existence of high bearing accuracy. If errors do exist, the best error mitigation technique is to acquire bearings on the subject HF surface wave signal from multiple, dispersed sites.

Combined surface and sky wave conditions create an extremely severe environment for accurate DF. Wave interference between the two waves produces rapid, deep amplitude fading with fade rates exceeding tens of Hz. The effective signal-to-noise ratio (SNR) is reduced, increasing direction-of-arrival errors. The fade rate may be comparable to the antenna scanning rate, negating optimal antenna sampling techniques. If the operating frequency is near the critical frequency, the ordinary wave component of the sky wave may disappear, leaving only the extraordinary component, which can experience tens of degrees of lateral displacement out of the great-circle path, thereby creating major bearing errors. The presence of both the ordinary and extraordinary sky wave components creates Faraday polarization rotation, which, in turn, produces polarization fading in a selective polarization antenna and instrumental errors in polarization-sensitive DF antennas such as loops.

Combined surface and sky wave conditions produce highly erroneous direction-of-arrival information. Effective mitigation methods for small-aperture DF systems are nonexistent. Fortunately, combined surface and sky wave conditions exist over only a relatively small part of the HF band from about 3 to 8 MHz.

The skip zone exists at frequencies above the critical frequency and at ranges beyond the effective surface wave distance. In the skip zone, no sky wave refracted signals are received; however, weak signals may be received by scatter mechanisms. Figure 4.9 illustrates ionospheric and ground scatter mechanisms. Clearly, the direction of arrival of scatter signals is inherently highly inaccurate. Accurate direction finding in the skip zone is problematic for any DF technique used; however, skip-zone signals exhibit unique amplitude-time and aural characteristics that afford identification.

The sky wave is the dominant propagation mode for HF communication and, hence, is of special interest for small-aperture DF. Due to the sky wave's importance to small-aperture DF, a separate section (Section 4.2.3) is devoted to it.

Another propagation mode deserves mention although it exists only under specific conditions. Propagation of EM waves from about 1 to 100 MHz in a forested environment can contain a lateral wave that travels along the tree tops. Characteristics have been defined by Tamir [20], and experimental investigations [21] have confirmed its existence. Figure 4.10 depicts the various wave configurations for a heavily forested environment. The forest, which is modeled as a dissipative slab, has two major interfaces: an air-forest and a forest-ground. For sky wave modes, the air-ionosphere interface is also pertinent. For dense forests, the lateral wave mode is the dominant "surface" wave mode because the others are highly attenuated. Because the lateral wave follows the contour of the tree tops, it can experience significant direction-of-arrival deviations from the great-circle path. The lateral wave phenomena has not been investigated extensively. A systematic, long-

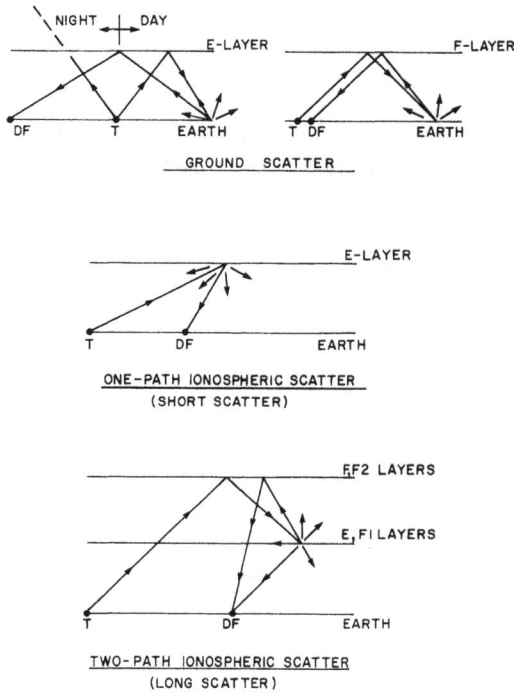

Figure 4.9 Ionosphere and ground scatter mechanisms.

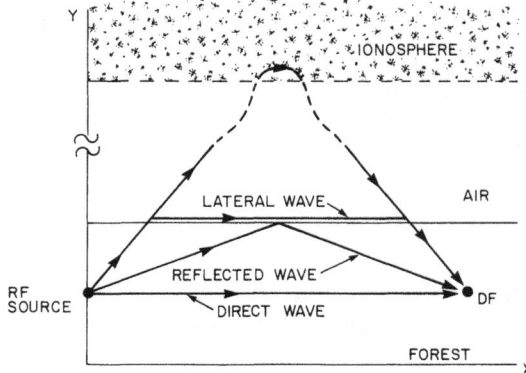

Figure 4.10 Wave components in a forested environment.

term investigation is needed to provide a reliable database for modeling and prediction.

Combined lateral and sky waves can create a severe DF environment that is similar to sky and ground wave interference. Further, the lateral wave is considerably depolarized with respect to other surface wave modes; therefore, depolarization-induced DOA errors may occur.

Very High Frequency (30 to 300 MHz)

Small-aperture VHF DF use extends from about 30 to 150 MHz. Above 150 MHz, directional medium-aperture DF antennas are practical—subject to host platform constraints. At VHF, the surface wave is useless for DF purposes, and DF must be performed on direct, reflected, or lateral waves.

Operational VHF DF experience indicates that the direction of arrival of the direct wave is generally accurate and lies on, or close to, the great-circle bearing. In this case, other error sources—usually environmental or instrumental—dominate the DOA error budget.

Reflection (and scatter) VHF propagation can deviate widely from the linear propagation path and produce erroneous direction-of-arrival conditions. Errors tend to increase as surface roughness increases. VHF DF in mountainous terrain is very unreliable—especially if the DF is not sited on high terrain. The same is true for VHF DF in built-up areas and heavily vegetated terrain where the lateral wave may experience significant deviation as it propagates across the tree tops.

Air-to-air and air-to-ground VHF DF must contend with combined direct and reflected wave interference that can produce rapid, deep amplitude fading [22]. VHF antennas elevated above the earth's surface propagate both a direct wave and an earth-reflected wave that combine to produce a lobing pattern. The lobing pattern is created by an in-phase and out-of-phase vector combination of the two waves. Movement of a DF system through the lobing pattern results in the imposition of lobe modulation (periodic signal amplitude fading) on the received signal (The presence of vegetation along the propagation path affects the lobing structure [23]). The maximum lobe modulation frequency f_{lm}, is given by

$$f_{lm} = (f)(v)/c$$

where f is the operating frequency, v is the relative velocity of the RF source and the DF system, and c is the velocity of light. For f in MHz and v in knots, f_{lm} is given by

$$f_{lm} = (0.00172)(f \text{ MHz})(v \text{ knots})$$

Figure 4.11 is a plot of f_{lm} *versus* relative velocity for operating frequencies of 50, 100, and 150 MHz. Fade rates of tens and hundreds of Hz are possible; rates of this magnitude can have an adverse effect on direction-finding performance and must be considered in VHF DF design and use.

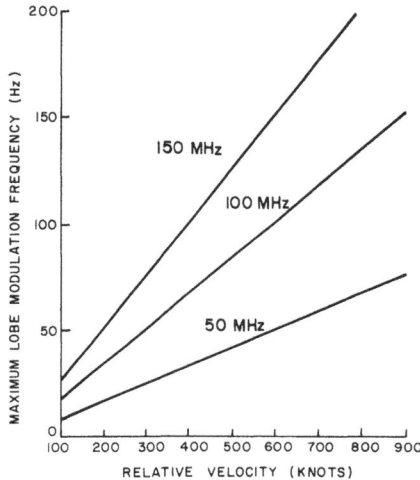

Figure 4.11 Lobe modulation frequency *versus* relative velocity.

4.2.3 Sky-Wave Direction-of-Arrival Errors

Range Sectors

Sky wave propagation is a major source of direction-of-arrival errors for small-aperture DF systems. Sky wave-induced errors involve a wide variety of mechanisms as was shown in Table 4.3. Specific mechanisms tend to dominate as a function of range from the RF source to the DF system. Therefore, DF errors show a strong correlation with range to the RF source. For purposes of analysis, three range sectors are generally used as shown in Figure 4.12. The three sectors (short, intermediate, and long) are delineated by elevation angle referenced to the zenith. Ranges are based on refraction from an F-layer assumed to be at a height of 270 km.

Short-Range Sector (0 to 200 km)

In the short-range sector, the sky wave has a steep angle of incidence to a DF site. The elevation angle is less than approximately 22°. For vertically polarized, *E*-field

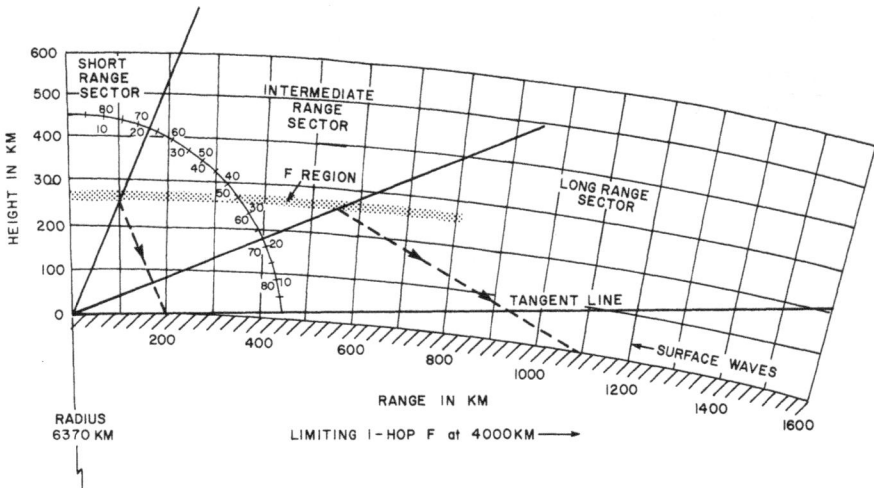

Figure 4.12 Sky wave sectors.

DF antennas, the sky wave response is reduced due to the steep elevation angle in the short-range sector. *H*-field (loop) antennas have an improved elevation angle response relative to *E*-field vertical elements, but the *H*-field elements are subject to polarization rotation errors at steep elevation angles.

The major source of short-range DOA errors is ionospheric tilts that produce lateral deviation from a linear propagation path. Figure 4.13 is the well-known Ross curve [24] depicting lateral deviation errors, as a variance function, plotted *versus* range for both day and night conditions. The Ross curve is based on several years of observations using azimuth angle-of-arrival data averaged over 5 minutes. Ross states that 5-minute bearing averaging delineates the slowly varying lateral deviation effects from the more rapid bearing fluctuations created by effects such as multipath and polarization rotation.

Bailey [25] extensively investigated lateral deviation DF errors. Figure 4.14 shows a plot by Bailey based on a 5-km lateral displacement of the ionospheric refraction point. Clearly, lateral deviation errors in the short-range sector (< 200 km) exceed several degrees, and errors can be as large as tens of degrees at very short ranges (tens of km). High-integrity sky wave direction finding at ranges of less than several hundred kilometers is generally agreed to be suspect unless sufficient ionosounde data are available to define the ionospheric tilt geometry and support compensation techniques [26].

Lateral deviation errors vary slowly as a function of time when compared to the errors introduced by polarization rotation and multipath wave interference;

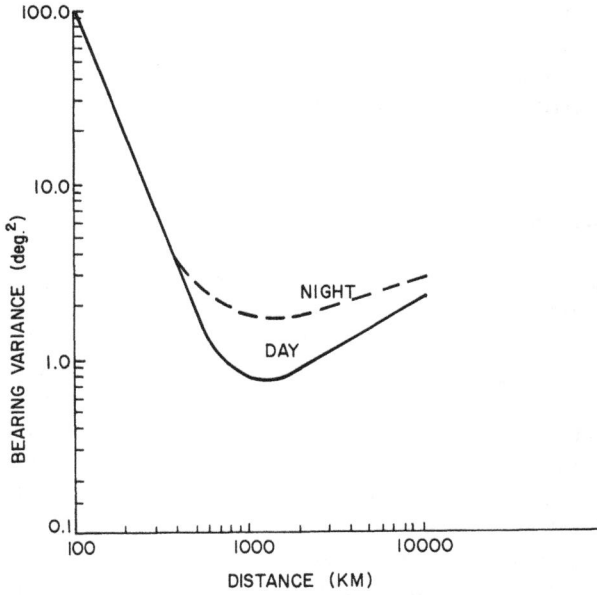

Figure 4.13 Lateral deviation *versus* distance: Ross curve [24].

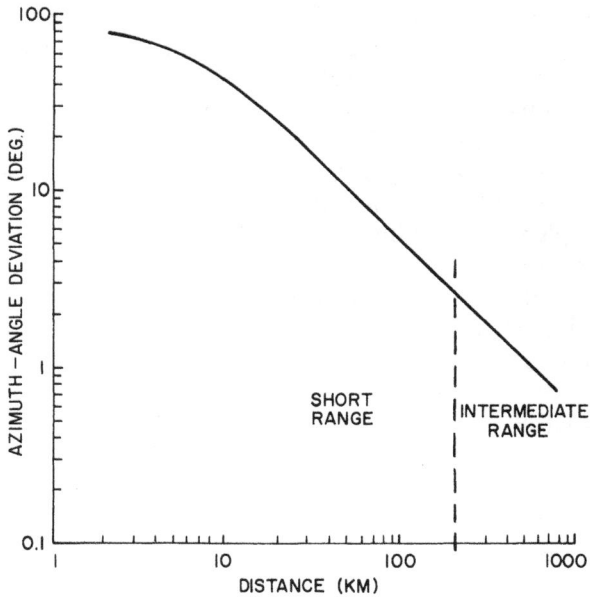

Figure 4.14 Lateral deviation *versus* distance: Bailey curve [25].

therefore, the potential for lateral deviation is high. However, even if compensation were applied to negate lateral deviation effects, large errors due to the other effects would remain. Polarization errors are especially high at steep elevation angles. Overall, small-aperture DF performance in the short-range sector is severely limited.

Intermediate-Range Sector (200 to 1100 km)

The intermediate-range sector exhibits a mix of sky wave error sources with no one source clearly dominant. The intermediate zone is the transitional region between the high-angle and low-angle sky wave regions; therefore, all sky wave-induced error sources are in evidence. Figures 4.13 and 4.14 show that lateral deviation errors decrease dramatically as the range increases to about 1000 km. In the intermediate-range sector, lateral deviation errors are comparable to other errors. Multipath wave interference errors become significant in the intermediate-range sector.

Multipath wave interference is created by multihop transmissions from the E- and F-layers, extraordinary and ordinary wave interaction, and ionospheric irregularities. These effects create multimode components and wave interaction. The results of investigations using special-purpose wide-aperture antennas (such as the Wullenweber array) have delineated the complex multimode structure of an ionospherically propagated signal [27]. These studies have indicated that the discrete modes in the incident multimode signal are not only of comparable amplitude but also arrive at different bearing angles. Further, the relative RF phases between the modes are time variable. The result of all of these variations is a corrugated phase front having a pattern that changes as a function of time.

Figure 4.15 (from Hayden [27]) illustrates the characteristics of multipath wave interference for a relatively simple two-component multipath field with a wave separation of 2α and relative amplitudes, h, of 0, 0.5, and 1.0. The two waves are assumed to be plane waves with their Poynting's vector parallel to the xy-plane. Hayden [27] points out that if the two waves were propagating in opposite directions ($\alpha = 90°$ in Figure 4.15), the maxima and minima of the wave interference pattern are spaced a half-wavelength apart. The minima-to-minima and maxima-to-maxima spacing is given by $\lambda/(2\sin\alpha)$; therefore, as α decreases from 90°, the spacing increases, but continues to lie in planes perpendicular to the direction of propagation. The equiphase surfaces become corrugated with the y-axis variations given by the following expression [27]:

$$Y = -[\lambda/(2\pi \sin\alpha)]\{\arctan[(1 + h/1 - h) \tan(2\pi X \cos\alpha/\lambda)]\}$$
$$- n\lambda/\sin\alpha \tag{4.1}$$

where n $= \cdots -3, -2, -1, 0, +1, +2, +3 \cdots$

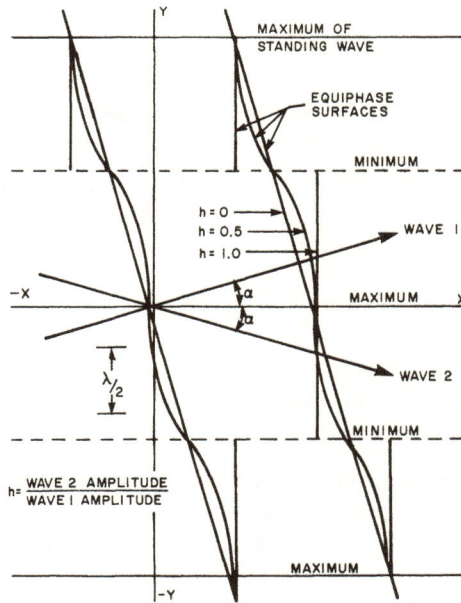

Figure 4.15 Multipath interference characteristics [27].

Equation (4.1) assumes a two-wave interference pattern with the two waves in phase at $x, y = 0$. For small α, the distance between the phase surface corrugations is greater than the dimensions of a small-aperture DF system. For example, with a typical α of 5°, the corrugation separation is about 5.5λ or 11 times the dimensions of a $\lambda/2$ aperture. For a small-aperture DF, the observed bearing at any instant will be normal to the phase surface at that point. Therefore, under multipath effects, small-aperture DF bearings are constantly shifting and fluctuating.

Small-aperture DF systems are unable to perform significant averaging across multipath corrugations; therefore, they provide erroneous direction-of-arrival information that can be as much as 90° in error. For example, Figure 4.16 shows a bearing error for a two-wave interference situation where the bearing error is plotted as a function of phase difference between the two waves. The larger amplitude, primary wave arrives at an angle $\alpha_p = 10°$, and the smaller amplitude, secondary wave arrives at an angle $\alpha_s = 50°$. The wave $2\alpha_p$ is 20°. The parameter h is the amplitude ratio referenced to the primary wave.

Figure 4.16 shows that the normal to the phase front fluctuates about the direction of the normal to the stronger, primary wave with the maximum error occurring at the out-of-phase (180°) condition. The error maxima occurs at a point outside the angle between the wave normals.

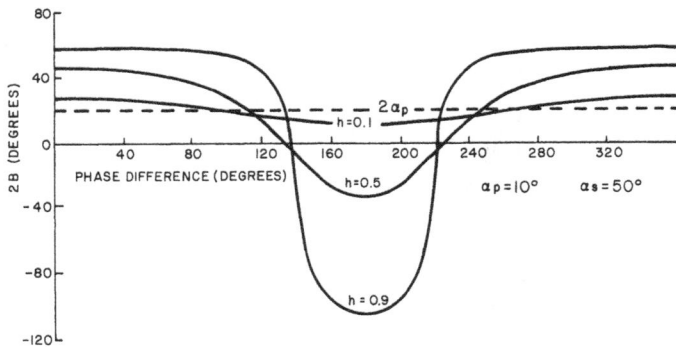

Figure 4.16 Bearing errors for a two-component field.

In many situations, ionospheric mechanisms create relative phase differences that are periodic, i.e., the phase difference increases monotonically and cycles through 2π radians in a periodic manner. This creates bearings that fluctuate back and forth across the direction of arrival of the stronger signal. When the signal amplitudes are equal, the bearing fluctuations cover 180°.

In the intermediate-range sector, ionospheric multipath is usually one-hop E- and F-layer modes; two- and three-hop paths exist, but relatively infrequently. Under these conditions, bearing time averaging is effective.

At sunset and sunrise, a sizable gradient exists in the electron density along an ionospheric path. This results in a "lumpy" ionosphere and leads to increased two- and three-hop multipath for each mode. Hence, bearing fluctuations become more erratic and less orderly. Bearing time averaging is less effective under these conditions.

Long-Range Sector (>1100 km)

This sector is characterized by (1) relatively low sky wave angles of elevation, (2) multiple-hop modes for both the E- and F-layers, and (3) an increase in lateral deviation errors. Low sky wave angles reduce polarization errors. Multiple-hop modes increase the variability of multipath interference and decrease the effectiveness of bearing time averaging. The increase in lateral deviation errors is due to the cumulative effects of multiple hops experiencing lateral deviation error offsets. The direction-of-arrival errors due to TIDs are greater for multiple-hop modes.

For distances greater than about 9000 km, long-path propagation can create direction-of-arrival errors approaching 180° especially when the long path is along either the terminator (the band separating day from night) or the nighttime portion of the earth.

4.3 ENVIRONMENTAL ERRORS

Small-aperture DF systems are often operated under conditions that are decidedly nonoptimal. The DF site usually contains reradiators, reflectors, surface obstructions, discontinuities in ground characteristics, and irregular host platform features. Environmental—or site errors—are introduced into the DF system by these effects because the DF system is an integral part of its immediate environs and reacts to adverse factors in the environs.

Environmental effects may be divided into three regions as a function of distance, r, from the DF system. The three regions are defined in the following sections.

4.3.1 Very Near Region

The very near region is in the immediate proximity of the DF system where $kr <$ 1 and k is the phase-propagation constant, $2\pi/\lambda$. Therefore, r is within $\lambda/2\pi$ of the DF system antenna. (If multiple antennas are used by the DF, the very near field is referenced to each antenna.)

DF antenna electrical performance (such as complex input impedance, current distribution, response pattern, and efficiency) is significantly affected by their host platforms. (DF antennas are sited on a wide variety of host platforms such as aircraft, boats, surface vehicles, elevated masts, and humans.) DF antennas are becoming smaller and more compact and, therefore, decoupling the DF antenna from the host platform is more difficult. Small-aperture DF antennas have robust near-field features, hence, when the DF antenna near field is close to the host platform, interaction between the antenna and platform occurs. (This includes transmission line interaction.) The objective then is to prevent the host platform from dominating the DF antenna performance and from functioning as the actual DF antenna.

Moment-of-method (MOM) computer codes are well suited for very-near-field DF coupling analysis. The Appendix discusses some of the MOM codes and their use for DF computations.

For ground-based DF systems, the ground *per se* may be in the very near field, and variations in ground dielectric constant and conductivity can cause direction-of-arrival errors. Subsurface features may also produce errors. Uniformity in the ground parameters is highly desirable—especially for systems using multiple, dispersed antenna elements. Mitigation techniques for nonuniform ground parameters at fixed DF sites are screening, counterpoises, chemical treatment of the ground, and elevation of the antenna. MOM analysis of ground effects is limited and requires considerable computational capability.

The human body is often the host platform for DF systems, especially DF systems used for search and rescue, wildlife tracking, assembly aids, and covert transmitter location. Computational techniques have been developed for analyzing antenna interactions with the human body [28, 29]; however, these techniques lack the maturity and efficiency of the moment-of-method computer codes.

4.3.2 Near Region

The near region extends from the very near region to a distance of 5 to 10 wavelengths from the DF system. For fixed station operation, this distance encompasses what is usually regarded as the "site area" where site calibration may possibly be performed and error mitigation practices applied. The host platform is generally an integral part of the near region.

The major error-producing effects in the near region are (1) point and extended reradiators and (2) earth and terrain irregularities. Most moment-of-method codes can handle the analysis of point reradiators; however, robust MOM codes are required for analyzing the effects of extended area reradiators and ground and terrain irregularities, especially for VHF DF operation.

Site reradiators are classified into point, linear, and sheet types depending on the physical configuration and size in wavelengths of the reradiator. MOM codes that match each type of reradiator are available and are especially useful at the HF and VHF bands.

The temporal characteristics of reradiators must be considered. For example, seasonal variations of vegetation and ground parameters can have a very significant impact on DF performance.

4.3.3 Far Region

The far region extends beyond the 5- to 10-wavelength range and includes the remote, outlying parts of the DF environs. Generally, the errors produced by the far-region effects are uncontrollable in the sense that calibration corrections are not feasible and computational efforts are not profitable. This is especially true if the DF system is a mobile ground-based unit. If the DF site is fixed, a long-term history of direction-of-arrival data on remote transmissions from known locations may provide some quantitative data on far-region error effects and may afford correction parameters.

Transitory reradiators, such as aircraft and helicopters, may occur in the far region. Large aircraft, such as commercial jumbo jets and military cargo planes, can produce significant specular and diffused reradiation over the entire HF and VHF region. Reradiation from smaller aircraft is most harmful in the upper HF and VHF bands.

Considerable attention has been given to the selection and preparation of DF sites that will result in reduced environmental errors. Requirements have been defined [30]; however, the requirements are generally not compatible with the use of small-aperture DF systems. The mobility and portability of small-aperture DF systems inhibit careful selection and preparation of sites. Small-aperture DF users are forced to accept decidedly nonoptimal siting conditions and maintain continual awareness of siting conditions in order to "grade" bearing data integrity.

The features of "good" and "poor" ground-based DF sites have emerged as a result of decades of experience. Table 4.4 delineates the features of a good site; Table 4.5 defines poor site characteristics. Numerous attempts have been made to define quantitatively the features of acceptable sites and delineate minimum distances to reradiators and irregularities. Success has been limited and restricted to specific siting situations. Fortunately, many small-aperture DF systems are either mobile or transportable and, hence, may benefit from operation at a diversity of sites with dissimilar siting errors. A homing DF system may be forced to traverse poor DF locations such as mountains, ravines, and overhead conductors. In these situations, an experienced DF operator may recognize poor conditions and avoid taking erroneous DF data.

Table 4.4
Characteristics of a "Good" Ground-Based DF Site

1. Sited on high, level, clear terrain.
2. Uniform, high ground conductivity and moisture content.
3. Void of features such as:
   - Extended aboveground conductors, e.g., power lines, telephone lines, antennas,
   - Extended subsurface conductors, e.g., pipelines, utility lines,
   - Wire fences,
   - Buildings,
   - Bridges,
   - Railroad tracks,
   - Water towers,
   - Chimney stacks,
   - Tall trees,
   - Rivers, streams, lakes, and
   - Coastlines.

Table 4.5
Characteristics of a "Poor" Ground-Based DF Site

1. Valley site in mountainous or hilly ravines.
2. Site near high cliffs or deep ravines.
3. Abrupt discontinuities in ground constants or features.
4. Rock or mineral outcroppings in site area.
5. Significant presence of any of the features listed under item 3 in Table 4.4.

In addition to good site selection and poor site recognition, other environmental error mitigation techniques can be adopted.

1. *Use H-field antennas: H*-field (loop) antennas are less susceptible to environmental coupling than *E*-field (monopole and dipole) antennas. Low terminal impedance and electrostatic shielding reduce capacitative coupling effects.

2. *Use balanced DF antennas and transmission lines:* Balanced DF antennas that are fed against ground have reduced susceptibility to environmental coupling. Transmission lines that are balanced and operated in a differential mode reduce the adverse effects of unbalanced capacitative coupling to the environment. Differentially fed, balanced loops and center-fed, ground-independent dipoles are essential for DF systems located on mobile platforms. Monopoles used at fixed ground sites should be differentially fed, and the ground effects should be stabilized by either a conducting counterpoise or chemical treatment.

3. *Maintain symmetry:* The maintenance of physical symmetry with respect to site reradiators and nonuniformities reduces coupling effects and provides error functions that are predictable and deterministic. For example, HF DF antennas symmetrically installed on an aircraft fuselage exhibit azimuthal quadrantal errors, i.e., the azimuth bearing error varies in a sinusoidal manner throughout 360° and contains two positive and two negative maxima. Quadrantal errors are generally well behaved and can be compensated for; however, physical symmetry is necessary for efficient calibration. Quadrantal errors also appear on other host platforms such as boats.

4. *Use elevated antennas:* For surface-based DF systems, elevation reduces environmental errors. For example, ship environments present severe DF siting conditions, and elevated antenna locations, such as mast tops, are essential. On ships, all frequency bands benefit from elevation; however, at HF, significant errors remain after elevation. These errors are due to hull and superstructure resonances that are most pronounced at HF [31]. VHF DF performance benefits greatly from antenna elevation. VHF DF systems with nonelevated antennas exhibit poor performance and are virtually useless in severe environments such as ships.

Quickly deployable, self-supporting masts are now available. (An example is the MAGIC MAST® series developed by GTE Sylvania. Loads up to 600 pounds can be elevated to 25 meters in 3 minutes. The vertical deflection in 40 mile/hour winds is 3°. The elevation of a DF antenna to 25 meters or greater should provide an interference-free very-near-field region for operating frequencies above about 5 MHz.)

4.4 INSTRUMENTAL ERRORS

Instrumental errors are created in the DF system as a result of either equipment imperfections or fundamental error mechanisms in the DF technique used.

Equipment imperfections encompass a wide variety of factors including the following:

1. Low SNR;
2. Amplitude and phase unbalance;
3. Time and frequency inaccuracies;
4. Hardware imperfections and aging;
5. Physical misalignment;
6. Digital processing and algorithm imprecision;
7. Calibration inaccuracies.

Low SNR errors are caused by an inability of the DF system to extract the maximum available SNR from the incident signal. Usually this is a consideration only in the upper HF and VHF bands where the receiver noise figure is noise limited internally; however, it is a major design consideration at lower frequencies when electrically small antennas are used.

Instrumental amplitude and phase unbalances are generally due to poor design or improper implementation. These error sources may be a function of numerous variables associated with signal characteristics (amplitude and bandwidth) and physical characteristics (temperature and aging). (Temperature and aging are insidious sources of amplitude and phase unbalance in cables, connectors, preamplifiers, *et cetera,* and are often difficult to monitor and assess.)

Time and frequency inaccuracies are direct causes of errors in time difference-of-arrival and doppler DF, respectively and are indirect sources of errors in DF systems using phase and amplitude DF techniques. Time base errors (TBEs) are particularly harmful to time interval measurements and digital processing based on signal sampling. Frequency offsets can create, for example, mistuned RF filters and result in mismatched phase shifts in multiple channel receivers.

The most serious hardware imperfections are those that (1) limit dynamic range and the full range of signal parameters and (2) change significantly with environmental effects and aging. Amplitude and phase stability and tracking (in multiple-channel receivers) are essential DF hardware requirements demanding special attention.

Instrumental errors are introduced by physical misalignment of the antenna elements from their optimal alignment. Misalignment may be due to either imprecise installation or the dynamics of the host platform (assuming the DF antenna system is not stabilized).

Imprecise digital processing of the basic DF information introduces instrumental errors. Analog-to-digital conversion can create quantization noise and time-sampling errors. In many cases, direction-of-arrival information is computed from basic digital DF data using software algorithms. A software algorithm is defined as a finite set of well-defined rules for the computation of a desired parameter in a finite number of steps. Algorithm imprecision is created by the necessary truncation of the algorithm to arrive at a finite set of rules and steps. Further, software computational algorithms are often based on assumptions about the statistical character of the signal or noise parameters. These assumptions may not be valid under certain conditions. In addition, further errors may be introduced when the software algorithms contain the mathematical approximations and interpolations required to reduce computer memory requirements and computational time.

Calibration inaccuracies are another source of instrumental error. Two major calibration error sources are radiated signal calibration and reference signal injection calibration. Radiated signal calibration uses a controlled target transmitter moved about on the circumference of a circle at least five wavelengths (5λ) distant from the DF site. The target transmitter is located at various surveyed azimuthal points. Ten-degree increments are often used. DF readings are taken and compared with the surveyed azimuth readings. Correction tables are prepared. Correction inaccuracies are introduced by imprecise azimuthal surveying and approximations in the correction table. The "gaps" created by the finite number of azimuthal reference points are smoothed or filled in by mathematical procedures such as interpolation or curve fitting. Both techniques are approximations with inherent inaccuracies.

The target transmitter must be far enough away from the DF antenna to minimize proximity effects. Ideally, the target transmitter should be located in the distant far-field (tens of wavelength); however, in most cases, this is impractical and the target transmitter must be operated at shorter distances. This introduces calibration errors as depicted by Figures 4.17 and 4.18, which were computed by Ross [32]. These figures plot DF antenna wavelength spacing *versus* target transmitter distance in wavelengths for a given calibration accuracy. They were computed specifically for Adcock DF systems with dispersed antennas; however, they are also generally applicable to other DF techniques using dispersed antennas. (For DF systems that use nondispersed antennas, the target transmitter distance should always exceed one wavelength to avoid near-field coupling effects.) For antenna elements spaced 0.25λ apart, Figures 4.17 and 4.18 show that the target transmitter distance should be 1.5λ and 2.4λ to maintain calibration errors equal to or less than 0.5 and 0.25 degrees, respectively. Note that radiated signal calibration deals with both site and instrumental errors. However, in many cases, the calibration site is well-prepared and conditioned; hence, the calibration data deal primarily with instrumental error definition.

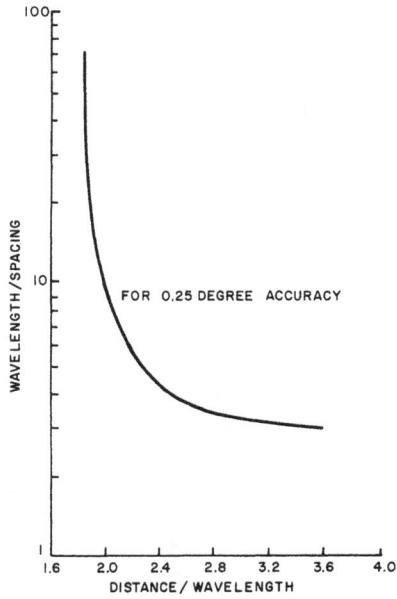

Figure 4.17 Calibration parameters for 0.25-degree accuracy [32].

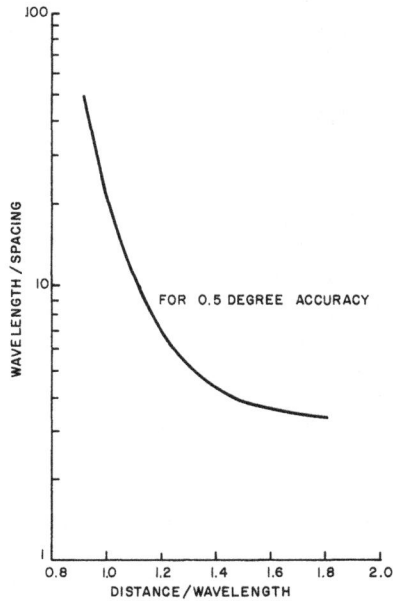

Figure 4.18 Calibration parameters for 0.5-degree accuracy [32].

Transmitter site reradiators are less harmful than DF receiver site reradiators. Assuming equal radiation efficiency and nominal conditions, transmitter site reradiations encompass a smaller field of view and subtend smaller off-path direction-of-arrival angles at the DF receiver site. Hence, on the average, transmitter site reradiations create less direction-of-arrival error deviation.

When using a surface-based target transmitter for DF azimuth calibration, elimination of or compensation for variations in the vertical, elevation plane pattern of the DF antenna are important. The effects of variations in the elevation pattern of the target transmitter may be reduced by maintaining a constant field strength incident on the DF antenna. This requires measuring field strength at the DF site as the target transmitter is moved in azimuth. The effects of DF antenna elevation pattern variations may be reduced by varying the height of the target transmitter antenna above the ground over a half-wavelength height difference (if practicable) and averaging DF angle-of-arrival measurements acquired at several heights.

Direct RF signal injection is commonly used in multiple-channel DF systems to measure the amplitude and phase unbalance between channels and provide correction parameters. RF signal injection is performed periodically (e.g., every minute) to obtain real-time correction parameters for DF techniques using either differential amplitude or phase measures to acquire direction-of-arrival information. The information obtained from RF signal injection significantly reduces instrumental errors created by amplitude and phase unbalance; however, some irreducible residual error usually remains due to inherent measurement inaccuracies.

4.5 OBSERVATIONAL ERRORS

Observational errors are those that occur at the operator interface, and are not necessarily operator errors *per se*. Observational errors are due primarily to the inability of the DF read-out or display to present uncorrupted information to the operator. Observational errors apply primarily to analog read-outs and displays. Digital read-outs are assumed to have negligible observational error unless we want to consider visual acuity as an error source.

Human factors and performance play a major role in establishing observational errors. For example, the aural-null method is used to measure the angle of arrival of an amplitude null DF system. The aural-null technique involves two major human factors: (1) the accuracy of the operator's eyes or ears in assessing the null point and reading the bearing and (2) the manual dexterity and expertise with which the null is manually swung around the amplitude null point, which indicates the angle-of-arrival.

The two major analog display techniques are the aural-null and the CRT display. For both analog techniques, observational errors depend on the signal chaacteristics and the condition of its display or read-out. Pertinent signal characteristics are as follows:

1. Signal-to-noise ratio;
2. Modulation type;
3. Signal duration;
4. Amplitude fluctuations (fading);
5. Interference or jamming.

Pertinent conditions are as follows:

1. Display linearity and balance;
2. Display size and layout;
3. Display scale visibility and readability;
4. Environmental effects on the operator;
5. Operator alertness and fatigue.

Observational errors received considerable attention from early DF practitioners. A comprehensive investigation of aural-null errors was conducted by Horner [33]. De Walden and Swallow [34] treated both aural-null and visual observational techniques.

In some application areas, direction-of-arrival readings are graded based on estimated quality. Generally, grading is divided into three classes: A, B, or C; 1, 2, or 3; *et cetera*. Grading is most useful for netted DF systems staffed by experienced operators. Direction-of-arrival grading has not been systematically defined or applied because its usefulness is problematic.

Remember, observational errors are not "operator" errors, *per se*. An alert, skilled operator is invaluable in reducing some observational errors as well as serving as an efficient "averaging" element capable of decreasing errors in the presence of DOA fluctuations and low signal-to-noise conditions. Further, operator expertise can help discern highly erroneous DOA conditions created by factors such as scatter propagation and site reradiation.

REFERENCES

1. Jay, Frank, ed., *"IEEE Standard Dictionary of Electrical and Electronic Terms,"* ANSI/IEEE Standard 100-1988, New York: Institute of Electrical and Electronic Engineers, 1988.
2. Jordan, E.C., and K.G. Balmain, *Electromagnetic Waves and Radiating Systems,* 2nd. Ed., Englewood Cliffs, NJ: Prentice-Hall Book Co., 1968, Chapter 16, pp. 608–655.
3. Wait, J.R., and J.A. Saxton, eds., *Advances in Radio Research: Electromagnetic Surface Waves,* London, England: Academic Press, 1964, pp. 157–217.
4. Terman, F.E., *Electronic and Radio Engineering,* New York: McGraw-Hill Book Co., 1955, Chapter 22, pp. 803–825.
5. Adams, J.E., *et al.,* "Measurement and Prediction of HF Ground Wave Propagation Over Irregular Inhomogeneous Terrain," NTIA Report 84-151, NTIS Access No. PB 85-110666, U.S.

Department of Commerce, National Telecommunications and Information Administration, July 1984.

6. Ott, R.H., "A New Method for Predicting HF Ground-Wave Attenuation Over Inhomogeneous Terrain," OT/TIS/TRER 7, NTIS Access No. AD 721179, U.S. Department of Commerce, Office of Telecommunications, January 1971.

7. Ott, R.H., *et al.,* "Ground-Wave Propagation Over Irregular Inhomogeneous Terrain: Comparison of Calculations and Measurements," NTIA Report 79-20, NTIS Access No. PB 298668/AS, U.S. Department of Commerce, National Telecommunication and Information Administration, May 1979.

8. CCIR, "Ground Wave Propagation Curves for Frequencies Between 10 kHz and 30 MHz, Recommendation No. 368-4, Propagation in Non-Ionized Media," Vol. V, CCIR XVth Plenary Assembly, Geneva, 1982.

9. CCIR, "Electrical Characteristics of the Surface of the Earth, Recommendation 527-1, Propagation in Non-Ionized Media," Vol. V, CCIR XVth Plenary Assembly, Geneva, 1982.

10. Hill, D.A., "HF Propagation Over Forested and Built-Up Terrain," NTIA Report No. 82-114, NTIS Access No. PB 83-194175, U.S. Department of Commerce, National Telecommunication and Information Administration, 1982.

11. Ehrman, L., A. Malaga, and F. Ziolkowski, "Communications Data Base Analysis for Military Operations in a Built-Up Area (MOBA/COBA)," Final Report for Period April 1977–August 1979, Report 15068.2, U.S. Army Contract DAAK29-77-C-0020, Lexington, MA: Signatron, Inc.

12. Malaga, A., and L. Ehrman, "HF MOBA Communications Study," Final Report for Period March 1979–July 1980, Report No. CORADCOM-79-0773-2, U.S. Army Contract DAAK80-79-C-0773, Lexington, MA: Signatron, Inc.

13. Davies, K., *Ionospheric Radio Propagation,* Washington DC: National Bureau of Standards, 1965, Monograph 80.

14. Bean, D.R., and E.J. Dutton, *Radio Meteorology,* Washington, DC: National Bureau of Standards, 1966, Monograph 92.

15. Cousins, M.D., "Direction Finding on Whistlers and Related VLF Signals," Technical Report 3432-2, SU-SEL-72-013, Office of Naval Research, Contract N00014-67-A-0112, Stanford, CA: Stanford Electronics Laboratory, Stanford University, 1972.

16. Horner, F., "Very-Low-Frequency and Direction-Finding," *Proc. IEE,* Vol. 104, Part B, No. 14, March 1957, pp. 73–80.

17. Malaga, A., and L. Ehrman, "HF MOBA Communications Study," Interim Report for Period March 1979–January 1980, U.S. Army Contract DAAK80-79-C-0773, Lexington, MA: Signatron, Inc.

18. *The Department of Transportation National Plan for Navigation, Proc. Nat. Aerospace Symp.* April 22–27, 1978, The Institute of Navigation, Appendix B, p. B-27.

19. Maloney, E.S., *Chapman Piloting,* 58th ed., New York: Hearst Marine Books, 1987, p. 546.

20. Tamir, T., "On Radio-Wave Propagation in Forest Environment," *IEEE Trans. Antennas Propagat.,* Vol. AP-15, No. 6, November 1967, pp. 806–817.

21. *Tropical Propagation Research,* Semiannual Report 7, Alexandria, VA: Jansky and Bailey Research and Engineering Department, Atlantic Research Corporation, July–December 1965.

22. Reed, H.R., and R.M. Russell, *Ultra High Frequency Propagation,* New York: John Wiley and Sons, Inc., 1953.

23. Chamberlin, K.A., "Investigation and Development of VHF Ground-Air Propagation Computer Modeling Including the Attenuating Effects of Forested Areas for Within Line-of-Sight Propagation Paths," Final Report AD A111719, Athens, OH: Avionics Engineering Center, Department of Electrical Engineering, Ohio University, March 1981.

24. Ross, W., "The Estimation of the Probable Accuracy of High-Frequency Radio Direction-Finding Bearings," *J. IEE,* Vol. 94, Part IIIA, March 1947, pp. 722–725.
25. Bailey, A.D., "HF Direction-of-Arrival Studies Over a Medium Range Path (452 km)," *Proc. Conf. HF Radio Propagation,* Urbana, IL: University of Illinois, June 2–4, 1969, pp. 13–30.
26. Johnson, R.L., *et al.,* "Short Time Scale Ionospheric Tilt Measurements for Lateral Deviation Compensation," *Proc. Conf. HF Radio Propagation,* Urbana, IL: University of Illinois, June 7–9, 1971.
27. Hayden, E.C., "Propagation Studies Using Direction-Finding Techniques," *J. Res. NBS,* Vol. 65D, No. 3, May–June 1961, pp. 197–212.
28. Chen, Kun-Mu, and D.P. Nyquist, "Coupling of Small Antennas with the Human Body," *Proc. Workshop Electrically Small Antennas,* Ft. Monmouth, NJ, May 6–7, 1976, pp. 171–176.
29. Neukomn, P.A., "Body-Mounted Antennas," Ph.D. Dissertation ETH No. 6413, Biomechanics Laboratory of the Swiss Federal Institute of Technology, Zurich, Switzerland, July 1979.
30. *Radio Direction Finding,* U.S. Army Technical Manual TM 11-476, Section XI, July 1947.
31. Crampton, C., "Naval Radio Direction Finding," *J. IEE,* Vol. 94, Part IIIA, March 1947, pp. 132–153.
32. Ross, W., "The Calibration of Four-Aerial Adcock Direction Finders," *Proc. IEE,* Wireless Section, Vol. 14, 1939, p. 299.
33. Horner, E., "Some Experiments on the Accuracy of Bearings Taken on an Aural-Null Direction Finder," Paper No. 868, Institute of Electrical Engineers, August 1949, pp. 359–365.
34. De Walden, S., and J.C. Swallow, "The Relative Merits of Presentations of Bearings by Aural-Null and Twin-Channel Cathode-Ray Direction Finders," *J. IEE,* Part III. Vol. 96, July 1949, pp. 307–320.

Chapter 5

SYSTEM LEVEL DESCRIPTIONS

5.1 FUNCTIONAL CATEGORIES

Small-aperture DF systems may be separated into four functional categories based on the fundamental operating principles presented in Chapter 2. The four basic categories are as follows:

- *Category I:* Systems using either direct or comparative amplitude response of the antenna subsystem for DOA information.
- *Category II:* Systems using the phase differential between disposed antenna elements with the phase differential converted to amplitude DOA information.
- *Category III:* Systems using the phase differential between disposed antenna elements for DOA information.
- *Category IV:* Systems using the time-of-arrival differential between disposed antenna elements for DOA information.

Each category may be subdivided into system classes based on antenna type, receiver channelization level, and DOA acquisition technique. The following material discusses each category, provides system-level descriptions, presents performance characteristics, and develops analysis methods.

5.2 CATEGORY I: AMPLITUDE RESPONSE

5.2.1 Classes

Figures 5.1 through 5.3 depict the three major system classes for small-aperture DF systems using direct and comparative amplitude response for DOA information. These systems use nondisposed antenna elements, which are predominantly electrically small *H*-field loop antennas.

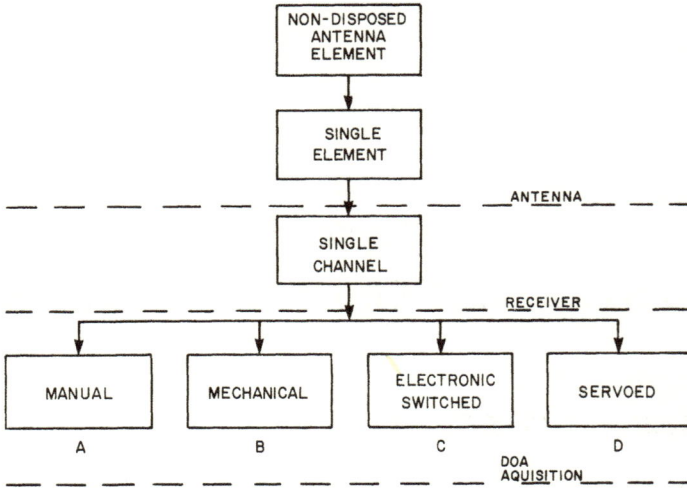

Figure 5.1 Amplitude response DF: single-element antenna–single-channel receiver class.

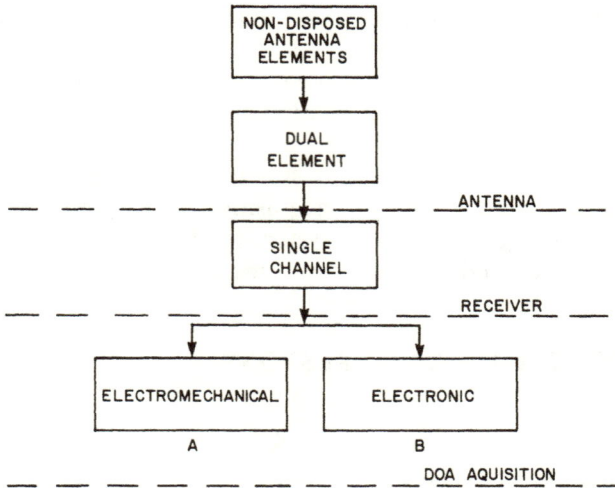

Figure 5.2 Amplitude response DF: dual-element antenna–single-channel receiver class.

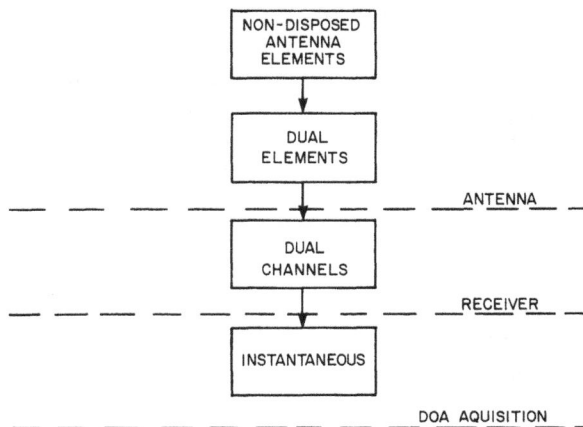

Figure 5.3 Amplitude response DF: dual-element antenna–dual-channel receiver class.

This category includes most of the earliest DF systems such as those using a single rotating loop, a servoed null-seeking loop, and goniometer-scanned crossed loops (the Bellini-Tosi system). The Watson-Watt crossed-loop, dual-channel receiver system is a Category I system. Many of these early DF systems are described in References [1–6]. Numerous Category I systems are in use today especially in applications involving cooperative emitters.

5.2.2 Single-Element Antenna–Single-Channel Receiver

Figure 5.1 delineates this class, which uses a single-loop antenna element and a single-channel receiver. (A sense antenna may also be used to resolve DOA ambiguity; however, the basic DOA information is acquired using a single antenna element.)

Type A: Manual Rotation

A single-loop antenna, with a separate sense antenna, is manually rotated about a vertical axis [4–7]. The azimuth bearing angle is determined by the operator using the aural-null (aural-minimum) method. A field strength meter is usually used for a visual indication of the amplitude null condition. A receiver beat frequency oscillator (BFO) is essential for acquiring aural nulls on unmodulated transmissions. Also, because a simple loop DF is often used for homing, an RF attenuator is needed to prevent overload; this switch is usually called a local–distant switch.

The first step of bearing acquisition is to activate the sense mode to obtain the cardioid pattern and then determine the approximate AOA by noting the signal maxima direction. The second step is to deactivate the cardioid path and use the more accurate figure-eight pattern null to acquire the bearing measure.

Strengths and limitations of the manual rotation technique are as follows:

Strengths

- Simplicity
- Portability
- Low power required
- Relatively high sensitivity (tuned loops used).

Limitations

- Very poor bearing accuracy on sky waves
- High bearing acquisition time
- Poor modulation tolerance
- Susceptible to environmental coupling (if tuned loop is used)
- Requires relatively high operator proficiency.

Type B: Mechanical Rotation

This type uses a simple vertical-loop antenna (with sense antenna) mechanically rotated about a vertical axis. Generally, the rotation is continuous at a maximum rate of several hundred revolutions per second. A CRT is often used to display the bearing information. Figure 5.4 shows a functional block diagram of a specific implementation [8].

In Figure 5.4, an ac motor rotates a balanced vertical loop and a sine–cosine angle encoder at a rate of ω_s. The rotating loop modulates its amplitude response pattern on to the incident RF signal. The modulated RF is then transmitted through RF slip rings to a superheterodyne receiver, where it is downconverted to an IF signal containing the ω_s pattern modulation and the azimuth bearing information.

The function of the post-IF circuitry is to detect and display the bearing information contained in the amplitude nulls of the scan modulation. The scan nulls are detected and converted to a baseband signal by a fast-acting AGC (FAGC), which is designed to detect the null pattern without degrading the null sharpness and depth. A conventional diode envelope detector simply reproduces the envelope of the scan modulated IF and, hence, degrades the null features in the diode square-

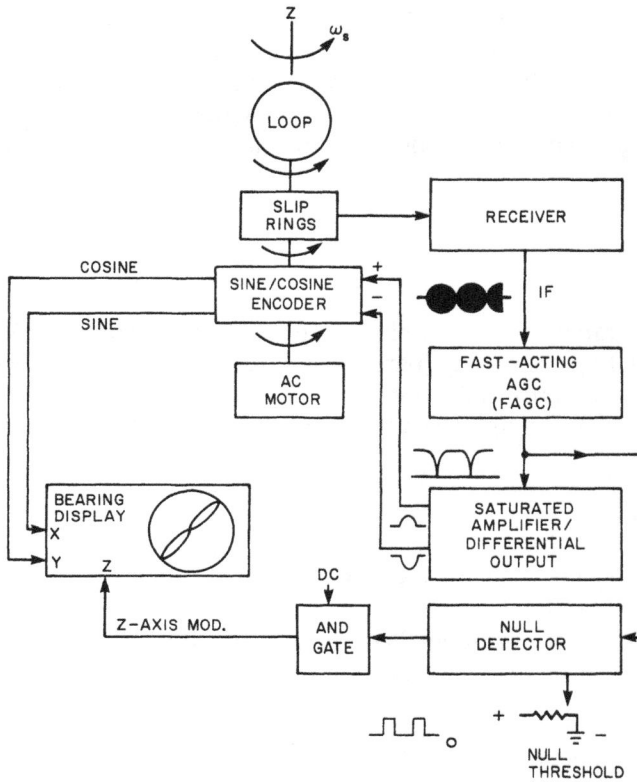

Figure 5.4 Amplitude response DF using mechanical antenna rotation.

law region. The FAGC maintains the IF signal at the diode detector out of the square-law region for a longer portion of the null excursion. The FAGC produces a sharper null and an enhanced bearing display. The FAGC output is applied to an amplifier that acts in a saturated state until the nulls occur. At the null minima, the amplifier clips, inverts, and amplifies the null region and provides differential outputs to a sine–cosine angle encoder, which is pulsed on by the differential "null" inputs.

Sine–cosine analogs of the loop azimuth angle are applied to the xy inputs of a high-persistency xyz oscilloscope. The xy inputs form a half-propeller pattern on the display with the tip of the display pointing to the bearing azimuth (or its recip-rocal). The next null paints the remaining portion of the propeller, hence, each null is separately displayed. An individual display of each null affords identification of erroneous bearings that exhibit asymmetrical null locations. (Erroneous bearings created by polarization rotation do not exhibit null asymmetry.) A null detector with an adjustable null detector threshold, provides z-axis (intensity) modulation to "pinpoint" the null bearings. The z-axis modulation and the high persistence of the oscilloscope provide a form of bearing averaging and, hence, error reduction.

The strengths and limitations of the mechanically rotated antenna system are as follows:

Strengths

- Relatively rapid bearing acquisition (compared to manual rotation)
- Relatively high sensitivity (narrowband, tuned loops may be used)
- Some inherent error reduction
- Less dependent on operator proficiency (compared to manual rotation).

Limitations

- Poor bearing accuracy on sky waves
- Poor modulation tolerance
- Susceptible to environmental coupling (if tuned loop antenna is used)
- High power is required (ac motor).

Type C: Electronic Switching

For Type C systems, antenna patterns are electronically switched, or lobed, to provide DOA information. Usually, lobed-pattern DF systems are used for either vehicle or personnel homing. In such systems, cardioid patterns are alternately switched left and right in the azimuth plane about the centerline of the host platform as illustrated by Figure 5.5 [6]. Pattern crossover occurs along the centerline affording a homing capability.

The incident RF signal is square-wave modulated by the pattern lobing. The modulation is detected by a receiver, converted to baseband (audio), and applied to the homing indicator. For the situation depicted in Figure 5.5, the indicator shows a right-of-centerline DOA. When the RF source is on the centerline, the indicator is centered, and a creditable homing bearing is present.

This homing technique is often used for VHF and UHF boat and aircraft homing for applications such as search and rescue and navigation to remote sites such as oil rigs, mining locations, *et cetera*. However, for these applications, the antenna system generally consists of disposed dipoles on a single baseline as discussed in Section 5.3.

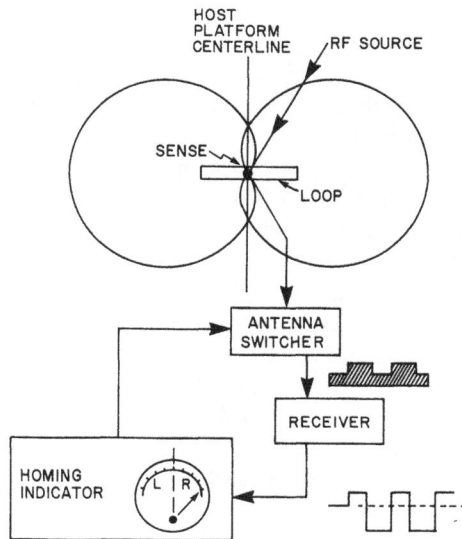

Figure 5.5 Amplitude response DF system using electronic switching.

The strengths and limitations of the lobing DF systems are as follows:

Strengths

- Simplicity
- Requires no special receiver (compatible with standard aircraft and marine receivers)
- Low power required
- Minimal operator proficiency
- Homing use reduces impact of low sensitivity.

Limitations

- Long bearing acquisition time
- Best suited for cooperative sources
- Poor accuracy on sky wave signals (when simple loop is used)
- Requires careful siting to preclude pattern distortion—especially at the cross-over point.

Type D: Servoed

The null of a cardioid antenna is used to servo a vertical loop to the null position and obtain a reading of azimuth AOA. Figure 5.6 shows a functional block diagram of a simple loop–sense antenna servoed DF system [9]. The cardioid pattern is formed by the loop–sense antennas and associated circuitry. An RF-balanced modulator is inserted in the RF circuitry to provide for the insertion of a low-frequency (Hz) error-sensing signal from the hertz reference source, which also provides a reference (REF) signal to one phase of the two-phase (2ϕ) ac motor. (Sixty hertz is often used as the error-sensing signal frequency.) The error-sensing signal modulated on the RF is detected, amplified, and applied to the variable (VAR) phase of the ac motor. The VAR phase either leads or lags the REF phase depending on the azimuthal direction of the RF signal relative to the loop null. The ac motor causes the loop antenna to rotate until a null condition is obtained. The phase lead or lag state determines the direction of rotation. The resolver and associated circuitry drive the bearing indicator to the bearing of the RF source. Single-loop servo DF systems are first-generation systems that have been followed by second-generation ADF systems such as those using dual loops and electronic scanning. The strengths and limitations are as follows:

Strengths

- Low operator proficiency required.

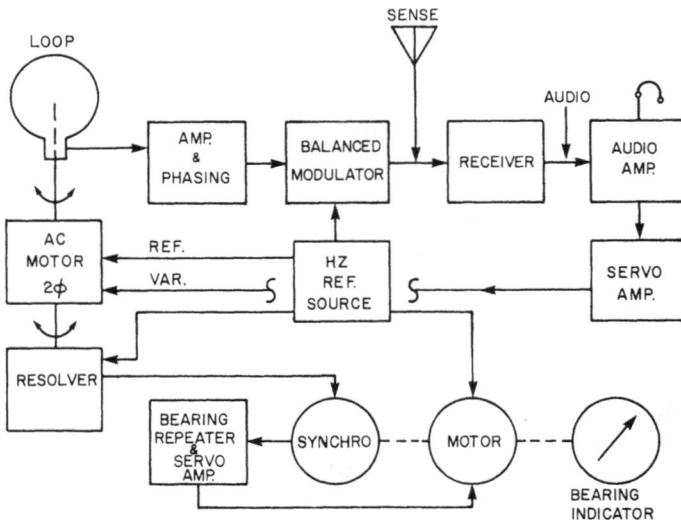

Figure 5.6 Amplitude-response servoed DF system.

Limitations

- Relatively large and heavy
- High power required
- High integrity antenna pattern required
- Very slow bearing acquisition (tens of seconds)
- Low modulation tolerance
- Poor accuracy on sky waves.

5.2.3   Dual-Element Antenna–Single-Channel Receiver

Figure 5.2 depicts the dual-element antenna–single-channel receiver class of amplitude response DF systems. The dual-element antennas are vertical, orthogonal, crossed loops with separate balanced outputs. The dual-element–single-channel class evolved from the need to improve significantly on the inadequate bearing acquisition time (e.g., tens of seconds) of the simple loop DF systems, which proved to be virtually useless on uncooperative transmissions of short duration (e.g., several seconds). Physical antenna movement was replaced by electromechanical and electronic scanning of the fixed crossed-loop configuration.

Type A: Electromechanical Scanning

This type of DF system uses a radio goniometer, which is defined [10] as an RF combining device used with a plurality of antennas so that the direction of maximum (or minimum) response may be rotated in azimuth without physically moving the antennas. The most common form of goniometer for MF and HF band usage is the inductive goniometer; capacitive goniometers generally operate above 100 MHz.

Figure 5.7 shows an inductive goniometer attached to dual, crossed-loop outputs. This is the Bellini-Tosi configuration. The two fixed windings of the goniometer are arrayed at right angles, and the internal, rotatable search coil can be rotated through 360°. The currents I_1 and I_2 in the fixed coils are assumed to be proportional to those in the corresponding loops 1 and 2. Therefore,

$$I_1 = I_{\max} \cos\phi \tag{5.1}$$

and

$$I_2 = I_{\max} \sin\phi \tag{5.2}$$

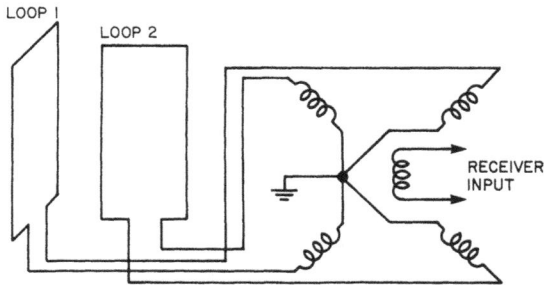

Figure 5.7 Inductive goniometer circuit.

where ϕ is the azimuth angle of arrival referenced to loop 1. The current I_3 induced in the movable coil is described by the following expression [5]:

$$I_3 \alpha (I_1 + jI_2) \alpha I_{max} (\cos\phi \, \cos\beta + j \, \sin\phi \, \sin\beta) \tag{5.3}$$

where β is the angle of the rotating coil relative to the plane of loop 1. The current I_3 has a maximum value when $\phi = \beta$ and a minimum value (null) when $\cos\phi = \sin\beta$. As the movable coil is rotated through 360°, its output varies from maximum and minimum two times per revolution, replicating the output of a single rotating loop. The goniometer's movable coil can be either rotated by hand for manual null determination or continuously rotated by a motor. In the latter case, a system such as the one shown in Figure 5.4 may be used to detect and display the bearings of the goniometer null. Pertinent strengths and limitations are as follows:

Strengths

• Relatively rapid bearing acquisition
• Efficient manual scanning for bulky antennas
• Relatively low power required.

Limitations

• Some loss in sensitivity in goniometer
• Poor accuracy on sky waves
• Low modulation tolerance
• Goniometer octantal errors possible.

Type B: Electronic Scanning

These systems use an electronic equivalent of the inductive (or capacitive) goniometer. The key device is the balanced modulator, which has an RF input and a low frequency (audio) input, which is used as the antenna scan. The scan modulation can be either analog or digital. Two balanced modulators are connected to two orthogonal antennas with figure-eight patterns (such as the patterns for crossed-loop antennas) and modulated by antenna scanning signals whose phase difference is $\pi/2$ radians. The two balanced modulator outputs are combined to form the electronic equivalent of the electromechanical goniometer [6, 11].

The principles of operation are illustrated by Figure 5.8. The azimuthal responses of the two orthogonal loops 1 and 2, with diameters much less than λ, are given by

$$E_1 = (k_1 \cos\phi - k_2 \sin\phi) \cos\omega_c t \tag{5.4}$$

and

$$E_2 = (k_1 \sin\phi + k_2 \cos\phi) \cos\omega_c t \tag{5.5}$$

where

E_1 = loop 1 output;
E_2 = loop 2 output;
ϕ = azimuth angle of the incident signal referenced to the plane of loop 1;
ω_c = $2\pi \times$ the frequency of the incident signal;
k_1, k_2 = functions of incident signal polarization and elevation angle of arrival, loop area and number of turns, and operating frequency.

The RF input to the receiver is given by

$$E_4 = f_1(t)E_1 + f_2(t)E_2 + E_3 \tag{5.6}$$

where $E_3 = k_3 \cos\omega_c t$ and k_3 is a function of the effective height of the sense antenna and coupling–combining losses.

Let $f_1(t)$ and $f_2(t)$ be sinusoids in time-quadrature where

$$f_1(t) = \cos\omega_s t = \cos 2\pi f_s t \tag{5.7}$$

and

$$f_2(t) = \sin\omega_s t = \sin 2\pi f_s t \tag{5.8}$$

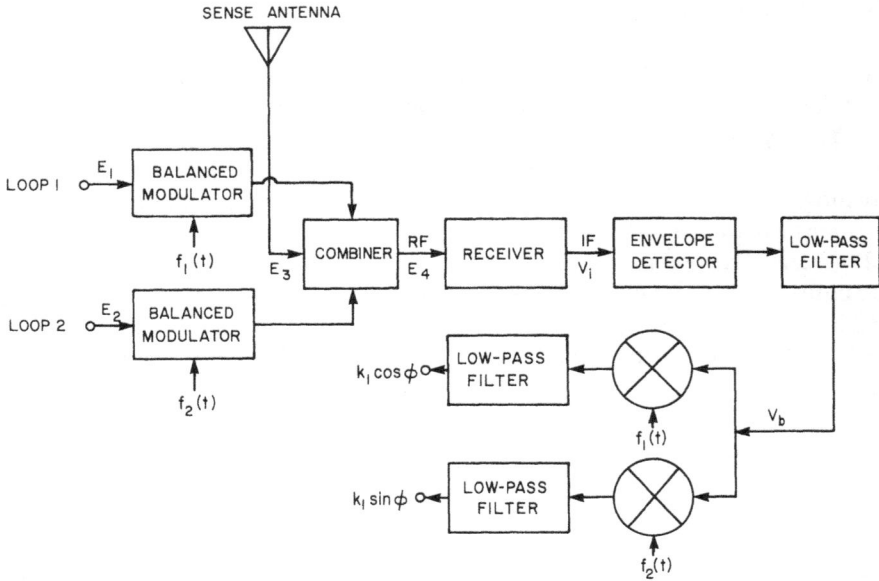

Figure 5.8 Amplitude response DF using electronic scanning.

The antenna scan frequency is f_s, which is typically chosen to be at several hundred hertz or less to avoid interfering with the information modulation on the signal of interest.

The receiver IF output is given by

$$V_i = [k_3 + k_1 \cos(\omega_s t - \phi) + k_2 \sin(\omega_s t - \phi)] \cos\omega_i t \qquad (5.9)$$

where $\omega_i = 2\pi \times$ the IF frequency and $|k_3| > |k_1 + k_2|$. For matched loops, identical RF modulators, precise quadrature scanning signals, and vertically polarized RF signals, k_2 is identically equal to zero.

After envelope detection and low-pass filtering of V_i, the resultant baseband signal, V_b, is synchronously detected with the antenna scanning frequencies $f_1(t)$ and $f_2(t)$. The synchronous detector outputs are low-pass filtered to recover the dc terms, which are $k_1 \cos\phi$ and $k_1 \sin\phi$. These dc terms may be further processed or displayed on a CRT to obtain a read-out of the azimuth angle of arrival ϕ with sense included.

The antenna scanning signals $f_1(t)$ and $f_2(t)$ can be set to different frequencies to provide for independent scanning of the response patterns.

The analog antenna scanning method has several major disadvantages. One, the parallel RF-balanced modulators and post-IF synchronous detectors must be matched over the entire scan modulation range and, two, because the balanced modulators are active analog circuits, they are susceptible to intermodulation, crossmodulation, and dynamic range degradations. These problems may be alleviated by digital switching of the balanced modulators with quadrature, binary $f_1(t)$ and $f_2(t)$ signals. The digital switching functions must be in time-quadrature, and the rate must be chosen to minimize interference to the information content of the signal. Most marine and aircraft MF automatic direction finders use the switch-scan technique. Systems of this type constitute a significant percentage of the small-aperture DF systems in use today; therefore, this class is discussed in more detail in Chapter 6.

Strengths and limitations of the electronic goniometer class are as follows:

Strengths

- Relatively rapid bearing acquisition
- Low power required
- Low weight and volume
- Low environmental susceptibility (untuned loops used).

Limitations

- Requires amplitude- and phase-balanced circuitry (this includes the loop antennas, which must be untuned to avoid severe tracking problems)
- Requires precise quadrature antenna scanning functions
- Susceptible to intermodulation and crossmodulation (analog circuits)
- Antenna scanning may interfere with signal information (This is not a major problem if the RF source is cooperative and its modulation is chosen to be compatible with the antenna scan modulation. If the signal is uncooperative, an unscanned intercept mode may be needed to extract signal information.)
- Low sensitivity (due to the use of untuned loops)
- Poor modulation tolerance (especially if the incident signal has spectral components at the antenna scan fundamental and harmonics)
- Poor accuracy on sky waves
- Relatively complex implementation.

The cumulative effects of these limitations imply poor performance on uncooperative signals of opportunity; therefore, most use has been limited to cooperative RF sources, such as radiobeacons, and operation under ground wave conditions only.

5.2.4 Dual-Element Antenna–Dual-Channel Receiver

Figure 5.3 details the dual-element antenna–dual-channel amplitude comparison class, which provides instantaneous DOA acquisition. An instantaneous amplitude comparison DF system is defined as any system that performs amplitude comparison of the signals received simultaneously by two or more antennas having substantially orthogonal directive patterns [10], such as patterns produced by crossed, vertical loops.

Figure 5.9 depicts two functional block diagrams based on the use of two crossed, vertical loops with loop 1 aligned on a north-south (NS) reference. (Watson-Watt and Herd [12, 13] used the direct RF amplification technique shown in Figure 5.9 for DF on electromagnetic sources of atmospheric origin [14]. This technique has become well known as the Watson-Watt system.)

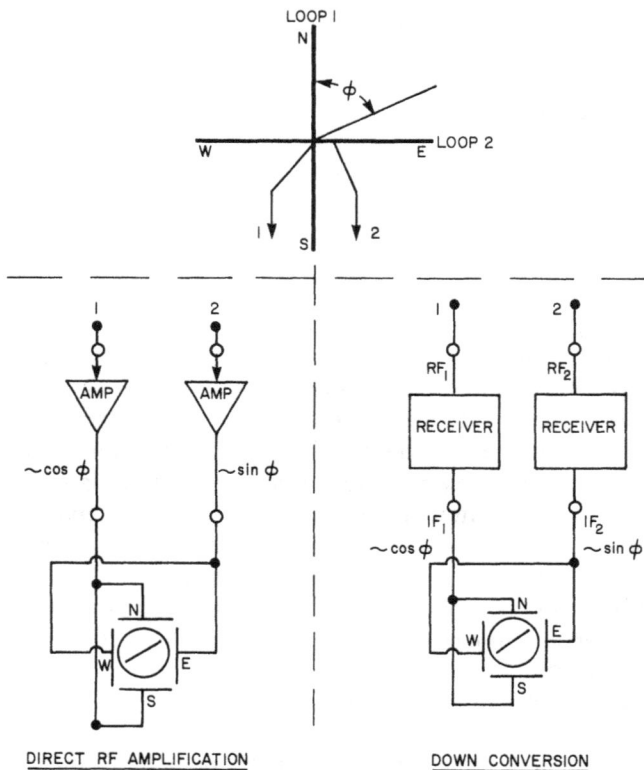

Figure 5.9 Amplitude comparison DF systems.

In Figure 5.9, each loop feeds a separate receiver or amplifier. The output of the NS (loop 1) receiver or amplifier is applied to the vertical deflection plates of a CRT; the east-west (EW) (loop 2) output goes to the horizontal plates. The slope of the in-phase Lissajou trace provides the signal bearing. Sense is not inherent in the dual-channel approach. A separate sense antenna with another amplifier or receiver channel may be required.

For vertical polarization, the azimuth bearing angle ϕ is given by

$$\tan\phi = V_{EW}/V_{NS} = \sin\phi/\cos\phi \tag{5.10}$$

or

$$\phi = \arctan[V_{EW}/V_{NS}] \tag{5.11}$$

where V_{EW} is the EW deflection plate voltage and V_{NS} is the NS deflection plate voltage.

The crossed-loop systems are susceptible to polarization errors. Consider the situation in which a single incident RF signal arrives from an azimuth angle ϕ and at an angle of incidence above the horizon of β. The incident signal contains both vertical (E_v) and horizontal (E_h) electric field components. The incident signal is experiencing polarization tilt or rotation. The loop responses are given by Eqs. (5.4) and (5.5), which may be normalized and expressed as

$$E_1 = (K/\cos\beta)(\cos\phi - X\sin\beta\sin\phi)(\cos\omega_c t) \tag{5.12}$$

and

$$E_2 = (K/\cos\beta)(\sin\phi + X\sin\beta\cos\phi)(\cos\omega_c t) \tag{5.13}$$

where K is a function of the loop characteristics and operating frequency and X is the ratio of E_h/E_v.

The measured azimuth angle ϕ_m is given by

$$\tan\phi_m = E_2/E_1 \approx \tan(\phi - \phi_e) \tag{5.14}$$

where ϕ is the true bearing and ϕ_e is the error in the measured bearing.

Equation (5.14) may be reduced to obtain

$$\phi_{\epsilon} = \arctan(X \sin\beta) \tag{5.15}$$

Equation 5.15 indicates a potential for large errors when the incident signal has a relatively large horizontal polarization component and a large angle of incidence.

This DF class is also susceptible to amplitude and phase mismatch between the two channels. Analysis shows [15] that amplitude and phase balance must be maintained to about 10% to contain bearing errors to less than 5°. This may be a severe requirement in the presence of cyclical temperature and humidity variations, vibration, and component aging. Further, the balance must apply across the spectrum of the incident signal. Asymmetrical amplification or phase shift of spectral components will result in bearing errors. Channel interchange switching and reference signal insertion to identify differential mismatches compromises the primary advantage, which is instantaneous bearing indication.

The Watson-Watt technique has never achieved widespread use. It was bypassed by phase interferometer techniques when the state of the art in amplitude- and phase-matched multiple-channel receivers became operationally acceptable. However, the need for truly instantaneous DF techniques has been revived by the advent of very short-duration burst and spread-spectrum transmissions, and interest in the Watson-Watt technique is increasing.

The major strengths and limitations of the Watson-Watt technique are as follows:

Strengths

- Instantaneous bearing indication
- Functionally simple
- Low power requirements
- No antenna scanning or switching required
- Portability
- Low operator proficiency needed.

Limitations

- Requires precision channel matching and tracking (for both amplitude and phase including the antenna)
- Low sensitivity (untuned loops used)
- Poor accuracy on sky wave signals

- Low modulation tolerance
- Relatively complex circuitry required (to achieve and maintain balance conditions).

5.2.5 Instrumental and Observational Errors

Instrumental Errors

The three dominant AOA instrumental error sources for amplitude-response DF systems are as follows:

1. Response pattern distortion;
2. Faraday rotation polarization errors [8, 16, 17];
3. Quadrantal errors.

Degradation in the response pattern results in null sharpness and directivity distortion and, hence, instrumental errors. The degradation is generally caused by extraneous signal pickup and unbalance. A reduction in null sharpness increases the null width and decreases the ratio of the null maxima-to-minima, n_r, which is related to the null azimuthal width, ϕ_n, by the expression

$$n_r = 114.6/\phi_n \tag{5.16}$$

where ϕ_n is the azimuthal width in degrees at the 3-dB values above the null. For example, a 1-degree null width requires a null ratio of 114.6.

Null degradation is due to signal pickup that is in quadrature with the desired loop signals. Null skewing is due to in-phase undesired antenna pickup. (For sensing purposes, the extraneous pickup is desired and controlled.) The null skewing error is called "reciprocal" error because the two nulls are displaced through equal and opposite angles as shown in Figure 5.10, which shows that the true bearing, ϕ_t, is one-half the angular distance, $2\phi_t$, between the true, displaced null and the reciprocal of the other null.

Errors introduced by incident sky wave signal depolarization are very significant and seriously compromise performance. Amplitude-response DF systems that use loop antennas derive DF accuracy from the existence of a correct, and constant, relationship between the direction of propagation and the polarization of the E-field vector. Referring to Figure 5.11, error-free polarization occurs when the incident E-field sky wave contains no horizontal component perpendicular to the direction of propagation. Any E-field polarization tilt, with $\theta < 90°$, creates horizontal components perpendicular to the direction of propagation and, hence, produces bearing errors.

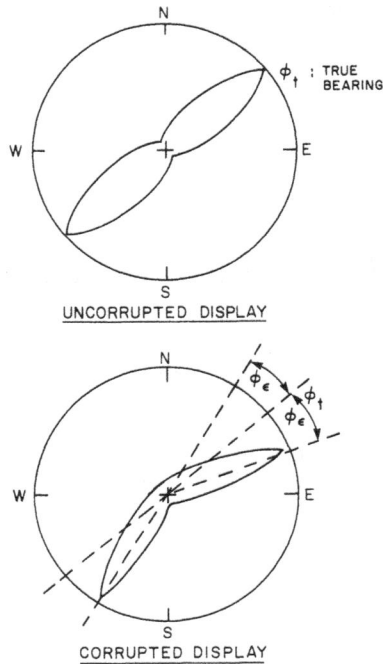

Figure 5.10 Bearing display showing reciprocal error effects.

Consider the case for a simple loop (Figure 5.11) in which polarization tilt introduces voltages in the loop that are not zero when the plane of the loop is perpendicular to the GCB. As the loop is rotated from the GCB, an amplitude null occurs but not at the correct bearing. Polarization bearing errors in a loop DF system are a function of the elevation angle of arrival and polarization tilt angle ψ. The general expression for the induced voltage, E_l, in an electrically small vertical loop located in free space is given by

$$E_l = KE(\sin\phi \cos\psi - \cos\phi \cos\theta \sin\psi) \qquad (5.17)$$

where

$K = (2\pi/\lambda)$ (area of loop) (number of turns);
ϕ = azimuth angle between the direction of propagation and the normal to the plane of the loop;
λ = wavelength;

ψ = polarization tilt angle; the angle between the E-field vector and the vertical plane normal to the plane of the loop;

θ = elevation angle between the direction of propagation and the plane of the loop;

E = the electric field strength in volts per meter.

If ϕ is not zero when a minimum value of E_l occurs, a bearing error results that is given by

$$\phi_\epsilon = \arctan(\cos\theta \tan\psi) \tag{5.18}$$

where ϕ_ϵ is the bearing error. [Equations (5.15) and (5.17) are identical with $K = \tan\psi$ and $\beta = 90° - \theta$.] Figure 5.12 shows the effect of depolarization on bearing

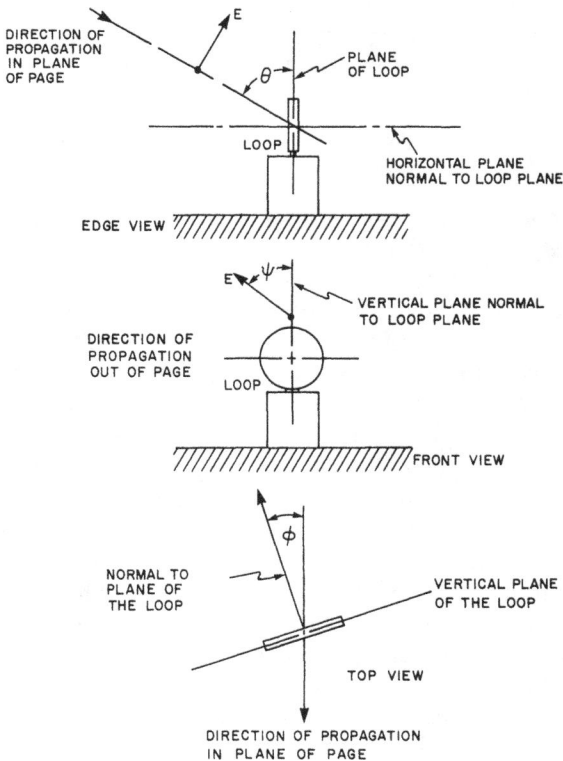

Figure 5.11 Diagram pertinent to polarization errors.

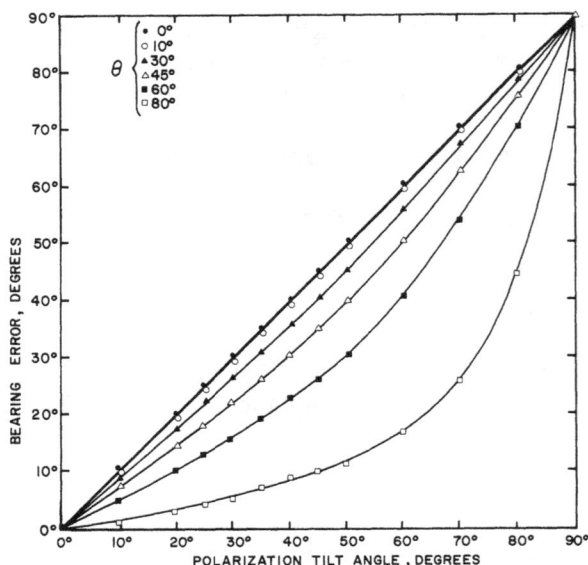

Figure 5.12 Bearing error *versus* polarization angle.

error and demonstrates that any nonzero value of ψ, with $\theta < 90°$, produces a bearing error. Because ϕ_c is a function of both θ and ψ, bearing accuracy can vary as θ varies even though ψ is constant.

Polarization tilt also decreases the sharpness and distinctiveness of the bearing null by decreasing the time rate of change of the induced voltage as the loop rotates through the null. This changes the slope of the null pattern and a broadening of the bearing display on a CRT. The relative change in slope is given by the function $\sqrt{1 - \sin^2\psi \sin^2\theta}$, which is plotted in Figure 5.13 for selected θ values. As depolarization increases, null sharpness degrades as elevation angle θ increases. Low elevation angles (e.g., 0 to 30°) show less than a 15% decrease in the time rate of change of induced voltage, and the degradation in bearing null sharpness is negligible. However, at high elevation angles (45 to 90°), null degradation is distinct and observable. Polarization errors are particularly prevalent when the sky wave signal is experiencing Faraday rotation and multipath effects are relatively insignificant. These conditions occur at operating frequencies near the MUF on one-hop paths. Also, these conditions are common in the short-range sector (Figure 4.12) where loop DF systems are often used. Figure 4.12 shows that F-layer propagation elevation angles in the short-range sector are generally less than 45°; therefore, relatively large polarization errors may be expected.

Faraday polarization rotation produces time-decorrelated fading between the vertical and horizontal components of the incident signal. Figure 5.14 shows a representative amplitude (vertical axis) *versus* time (horizontal axis) plot of the vertical and horizontal components of Faraday polarization rotation fading. The asynchronous nature of the fading is evident in that the vertical component nulls and peaks occur at the horizontal component peaks and nulls, respectively. (The transmission distance was 482 km at the path MUF.)

For Faraday polarization rotation, θ remains relatively constant and ψ varies with time at a relatively uniform rate [18]. Figure 5.15 shows a plot of $d\phi/d\psi$ *versus* ψ for selected θ values. This plot shows that the bearing display on a CRT rotates in a nonlinear manner as ψ rotates uniformly with time. The time rate of change of ϕ is smallest when ψ is vertical (0 or 180°), and this point corresponds to a minimum bearing error. Conversely, bearing errors are largest when ψ is 90 or 270°. At the larger elevation angles (90° to about 45°), variations in bearing swing rate are very pronounced as ψ rotates through 2π radians. Signals arriving at the smaller elevation angles produce less bearing swing rate variation as ψ rotates through 2π radians. Figure 5.15 indicates that the bearing swing rate is a function of elevation

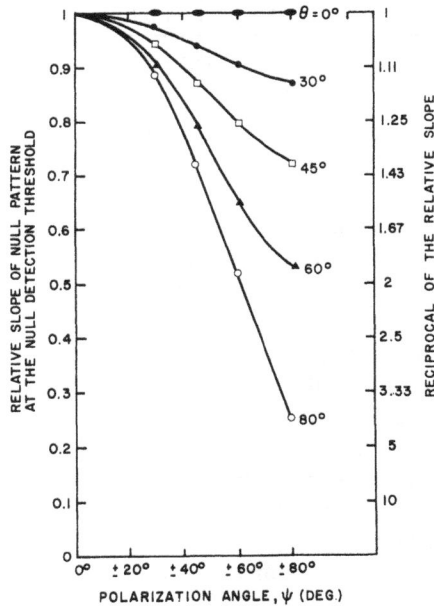

Figure 5.13 Relative slope of the null pattern as a function of elevation angle and polarization tilt angle.

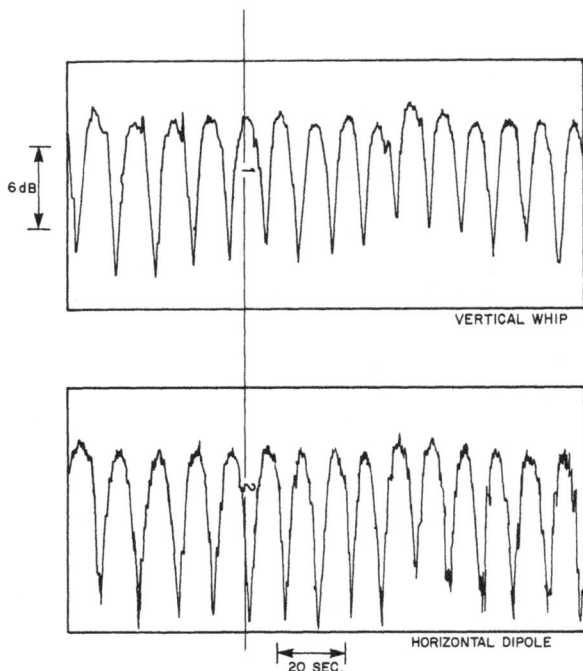

Figure 5.14 Representative plots of polarization fading.

angle θ. Reference [8] has demonstrated that the elevation angle θ can be approximated by measuring bearing swing rate parameters.

To summarize, bearing polarization errors can be as large as $\pm 90°$ from the GCB. Inaccurate bearings are evidenced by swinging bearings and display broadening, with the slowest swing rate and sharpest bearing occurring on or near the GCB. As the elevation angle decreases, the swing rate and bearing display become more uniform over a 2π radian rotation of polarization.

Experienced, proficient operators can discriminate between accurate and inaccurate bearings based on null sharpness and swing rate, especially at the larger elevation angles; however, for θ values decreasing below about $45°$, discrimination based on bearing features is more difficult.

Rotating loop DF performance is degraded when the signal of interest modulation is either interrupted carrier (ICW) or carrier-less single sideband (SSB). The "off" periods of signals with interrupted modulation may appear similar to authentic bearing nulls on a CRT display. This is especially true if the signal interruption rate is comparable to the DF antenna scanning rate. The clutter formed creates a very difficult bearing display pattern for the operator to interpret and, hence, degrades response time and accuracy.

Polarization error reduction techniques have been investigated. Two basic approaches are (1) compensation for the error-producing components and (2) the elimination of inaccurate bearing displays and read-outs [8]. Some compensation error-reduction techniques [19–21] have focused on the use of a horizontal dipole, rotating in synchronism with the loop, to provide a neutralizing voltage for the voltage induced in the loop by the undesired horizontal component of the incident signal. Effectiveness is a function of how well the compensation signal tracks the undesired component in magnitude and phase. Compensation adjustment depends on the height above the ground, ground-reflection conditions, operating frequency, elevation angle of arrival, and the relative effective electrical height and orientation of the loop and dipole. Effectiveness is a sensitive function of height above the ground in wavelengths. The best performance occurs at heights that are small compared to a wavelength (e.g., $\lambda/10$). (This technique is not compatible with aircraft operation.)

Another compensation technique [22] involves the use of a single, tilted rotating loop as depicted in Figure 5.16. Tilting is performed by mechanically position-

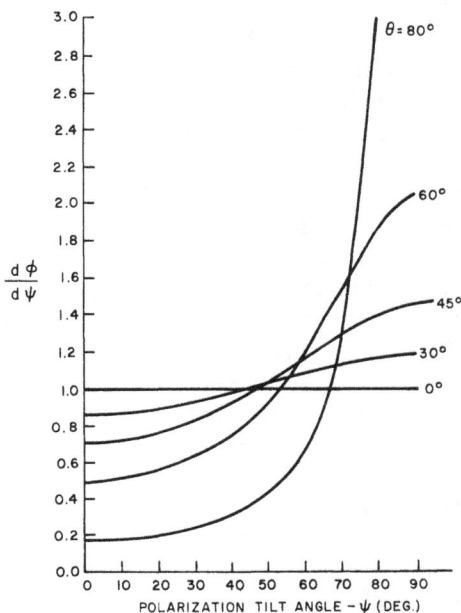

Figure 5.15 Derivative of bearing error as a function of polarization tilt angle.

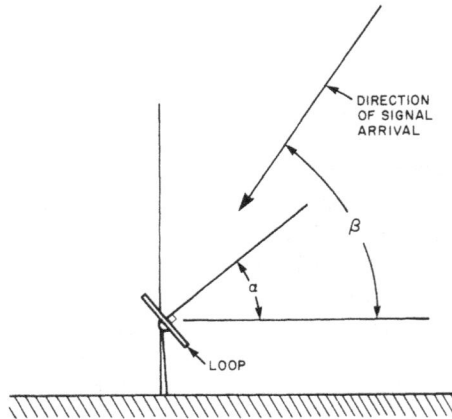

Figure 5.16 Tilted loop technique.

ing the plane of the loop at a tilt angle α relative to the horizontal while maintaining rotation about a vertical axis. The tilt angle α is variable from 0 to 90°. The signal is incident at an angle β relative to the horizontal. In the presence of polarization rotation, the tilting produces distinct and interpretable differences between accurate and erroneous CRT bearing displays. When $\alpha \approx \beta$, the differences in the display features are pronounced, and elevation angle of arrival can be approximated. This technique requires a relatively long transmission duration due to the time consumed in mechanically tilting the loop and observing the display over a polarization rotating cycle. Field tests demonstrated that the tilted loop technique provides considerable bearing error reduction relative to a simple rotating loop technique.

For rotating loop DF systems, the elimination of inaccurate bearing displays created by polarization rotation may be accomplished by displaying only those bearings occurring at or near the maxima of the vertical E-field component at the crest of a polarization fade cycle. This is accomplished by using display "unblanking" based on fade crest detection. A dual-channel receiver is used. (Precise amplitude and phase tracking is not necessary.) One channel is the conventional loop DF receiver; the other channel receives and detects the output of a vertical monopole sensing the vertical E-field component. The detected amplitude-time variation of the vertical E-field component is processed and used for display unblanking. Fade crest unblanking was developed and field tested [8]; it proved to be an effective error-reduction technique. Here again, relatively long transmissions are required to achieve the most effective performance.

A dual-channel DF receiver provides a method for reducing bearing display clutter created by interrupted transmissions such as ICW and SSB. The output of a vertical monopole receiver channel is used as a gating signal to gate out the false amplitude nulls created by carrier interruptions and pass the true bearing nulls. Clutter reduction is based on the fact that the bearing nulls occur only when the vertical monopole channel output is high and the loop DF channel is low, i.e., at a bearing null. The vertical monopole channel output does not contain the loop amplitude scan modulation but it does contain the interrupted carrier "nulls."

Quadrantal errors are azimuthal bearing errors that vary in a sinusoidal manner over 360° with two positive and two negative maxima [10]. Quadrantal errors were originally identified with host platform effects. (Loop DF systems sited on boats and aircraft usually exhibit quadrantal errors.) However, quadrantal errors are caused by any factor that creates amplitude unbalance in the DF system. Quadrantal errors are especially evident in crossed antenna systems—either crossed loops or crossed Adcock arrays. For example, the outputs of two crossed antennas are given by

$$E_1 = k \sin\phi \tag{5.19}$$

and

$$E_2 = k \cos\phi \tag{5.20}$$

Let E_1 and E_2 experience a relative amplitude unbalance of α such that

$$E_1 = k(1 + \alpha) \sin\phi \tag{5.21}$$

and

$$E_2 = k(1 - \alpha) \cos\phi \tag{5.22}$$

If the DF technique uses the relationship

$$\tan\phi = \sin\phi/\cos\phi \tag{5.23}$$

to measure ϕ, then the bearing angle ϕ_m measured with amplitude unbalance is given by

$$\tan\phi_m = [(1 + \alpha)/(1 - \alpha)] \tan\phi \tag{5.24}$$

The bearing angle error is given by

$$\phi_\epsilon = (\phi_m - \phi) = \alpha \sin 2\phi \tag{5.25}$$

where ϕ is in radians, α is a fractional expression, and ϕ_ϵ has a positive or negative maximum in the middle of each quadrant. Phase unbalance between the crossed antenna elements does not create a bearing error but does create quadrantal variations in null width.

Quadrantal errors also occur if the crossed elements or baselines are physically misaligned and do not cross at a 90° angle. If α is the angular azimuthal misalignment error and is positive for one element and negative for the other, the quadrantal error is

$$\phi_e = \alpha \cos2\phi \tag{5.26}$$

The pitch and roll of an airborne platform can produce disorientation errors in crossed-loop systems. Hence, ADFs are susceptible to disorientation errors. One loop (loop 1) is aligned along the aircraft fore-aft centerline and the other loop (loop 2) is aligned along the horizontal perpendicular to the fore-aft centerline. As the aircraft pitches and rolls, the crossed-loop responses, given by Eqs. (5.19) and (5.20), become

$$E_1' = k \cos\alpha_p \sin\phi \tag{5.27}$$

and

$$E_2' = k \cos\alpha_r \cos\phi \tag{5.28}$$

where α_p is the pitch angle and α_r is the roll angle. The elevation angle is assumed to be 90°. The bearing to the RF source is measured as

$$\begin{aligned} \phi' &= \arctan(E_1'/E_2') \\ &= \arctan[(\cos\alpha_p/\cos\alpha_r)(\tan\phi)] \end{aligned} \tag{5.29}$$

Maximum errors occur in each quadrant at angles of $n \times 45°$ where $n = 1$, 3, 5, and 7. No errors occur along the fore-aft centerline or the perpendicular to the centerline. Disorientation errors can be significant. For example, with a 20° pitch or roll, the error is approximately 1.8°. Whenever possible, all ADF bearings should be obtained during stable straight and level host platform conditions.

Quadrantal errors may be reduced by careful attention to circuitry amplitude balance and physical alignment; however, quadrantal errors due to the environs and host platform effects may be irreducible and must be measured and treated as site errors.

Quadrantal errors of DF systems using electronic goniometers and multichannel tracking receivers are often due to circuitry unbalance and parameter drift. Balanced modulator drift is especially harmful. Periodic calibration is required using either a local RF source or direct RF signal injection. Quadrantal errors are a function of RF frequency; therefore, calibration should be performed across the entire operating range.

Observational Errors

Angle-of-arrival readouts for amplitude DF systems use aural-null and visual bearing display techniques, which include CRTs, analog electronic and electromechanical meters, and digital indicators.

Aural-null errors are a function of both the sharpness of the null and operator skill. A normal human ear can detect an audio change of about 3 dB, and aural-null DF systems are designed using this criteria. Accurate null definition is a function of operator expertise in discerning this 3-dB change in signal level, which may be embedded in noise and interference. If the aural-null technique is augmented by visual methods, such as a signal strength meter, accuracy increases.

Visual observational errors are a major function of display scales and their readability. Azimuth scales that require extensive interpolation introduce observational errors. Also, nonlinear scales (created to compensate for quadrantal errors, for example) enhance observational errors. Inexperienced operators tend to group bearings to the nearest major graduation on the scale.

5.3 CATEGORY II: PHASE DIFFERENTIAL-TO-AMPLITUDE RESPONSE

5.3.1 Functional Approach

Category II systems are similar to Category I amplitude-response systems in that the primary direction-of-arrival information is obtained from signal amplitude; however, for Category II systems, amplitude information is obtained from the phase delay across the baseline of disposed antenna elements as shown in Figure 2.6. For antenna spacing, d, equal to or less than about 0.1λ, the azimuth response is approximately sinusoidal with a figure-eight pattern as depicted by Figure 5.17, top. Dual, crossed baselines produce the well-known orthogonal figure-eight response as shown by Figure 5.17, bottom.

The basic phase-to-amplitude antenna consists of two vertical elements connected as shown in Figure 5.17, top. (This is the classical H-Adcock antenna patented by F. Adcock in 1919.) A crossed-baseline configuration is also shown in Figure 5.17, bottom. The crossed arrangement of the transmission lines performs two functions: (1) π-radian (180°) phasing between the antenna elements to develop the proper differential across the baseline and (2) cancellation of voltages induced in the horizontal transmission lines by horizontal polarization components of sky waves.

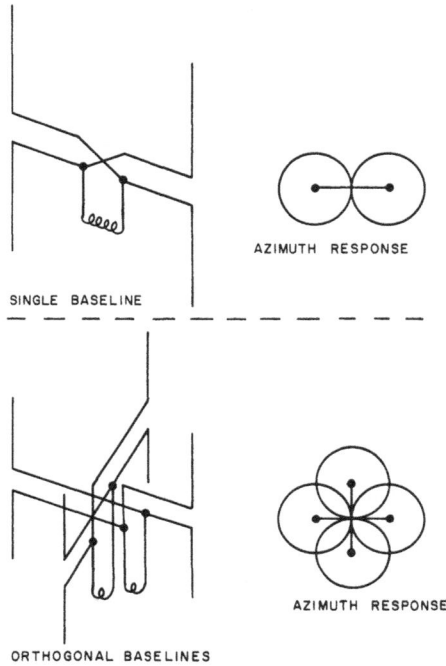

Figure 5.17 Adcock antenna configurations.

The response patterns of the Adcock configurations are identical to those using loop antennas for vertically polarized signals, and the Adcock configuration can function as a direct replacement for the loop configurations. The major advantage of the Adcock antenna is its relative immunity to polarization errors.

Both E- and H-field elements may be used for the phase-to-amplitude technique. (Folded dipoles are frequently used for the H-Adcock elements.) Figure 5.18 depicts several antenna configurations that have been used in operational systems. The U-Adcock is primarily used for nonrotating, crossed-baseline systems at fixed sites in the MF and HF bands. Uniform ground constants and horizontal member shielding are essential for satisfactory operation. The H-Adcock is typically elevated aboveground, and is generally used in the upper HF and VHF bands. H-Adcock antennas are widely used for marine VHF RDF for both position location and search and rescue. The spaced coaxial loop variant is relatively compact and compatible with rapid (e.g., 300 revolutions/minute) mechanical rotation. Also, compared to E-field elements, loop elements have better elevation angle response near the zenith and are less susceptible to environmental coupling effects. The spaced coaxial loop antenna has undergone considerable development for ground-based and shipborne HF and VHF applications [23–29].

The 180° DOA ambiguity of the Adcock antenna's symmetrical response pattern may be resolved by several methods such as: (1) add a separate omnidirectional vertical sense antenna at the phase center of the baseline(s), (2) introduce deliberate unbalance in the antenna elements or transmission lines to alter the response pattern in a deterministic manner, or (3) combine the antenna element outputs to obtain a separate omnidirectional response.

The conventional Adcock antenna uses a balanced RF transformer as an output element. When the RF transformer is replaced with an RF sum-and-difference coupler, another important class of DF emerges. Disposed antenna elements arrayed along baselines may be differenced (Δ) and summed (Σ) to obtain simultaneous directional and omnidirectional response outputs. When used with dual-

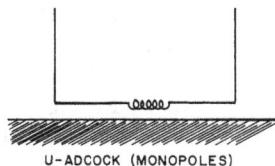

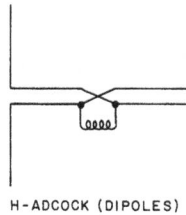

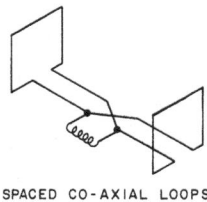

Figure 5.18 Adcock antenna types.

channel receivers, the sum-and-difference (Σ/Δ) technique provides improved performance over the basic Adcock technique.

Other Category II types are the probed line, phase compensation, and enhanced phase-to-amplitude conversion techniques.

5.3.2 Classes

Figures 5.19 through 5.21 delineate the various classes of DF techniques that convert the phase differential between disposed antenna elements to amplitude to acquire signal direction of arrival.

5.3.3 Single-Baseline Antenna–Single-Channel Receiver

Figure 5.19 breaks out the class using two antenna elements disposed on a single baseline and a single-channel receiver. (Ambiguity resolution may require another antenna located at the phase center of the baseline.) The Adcock antenna technique is used for all four of the types shown in Figure 5.19, and, because the Adcock antenna functions as a "direct" replacement for the loop configurations of Category I systems, the systems described in Section 5.2 may also be used with Adcock antennas.

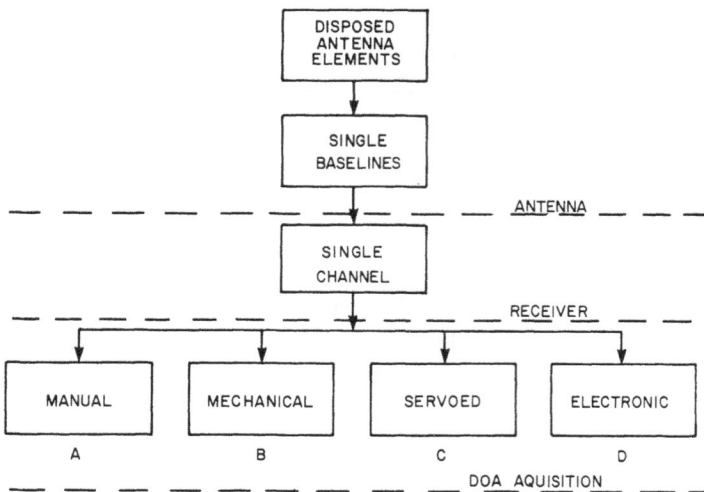

Figure 5.19 Phase differential-to-amplitude DF: single-baseline antenna–single-channel receiver class.

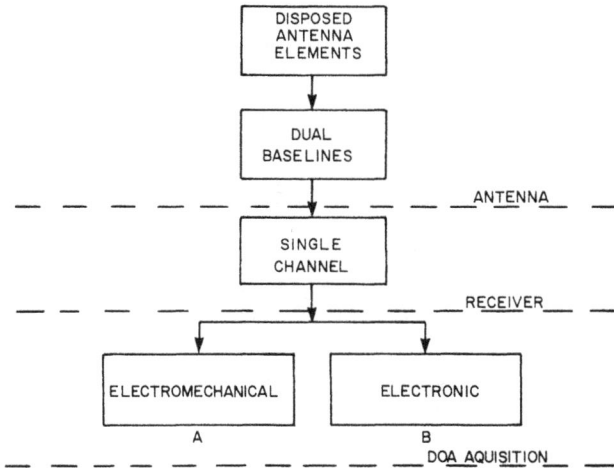

Figure 5.20 Phase differential-to-amplitude DF: dual-baseline antenna–single-channel receiver class.

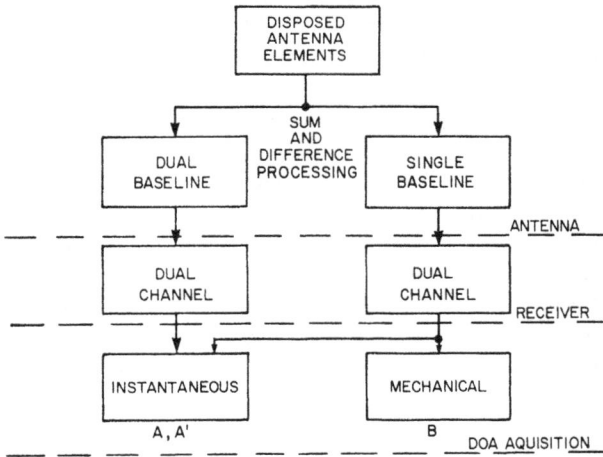

Figure 5.21 Phase differential-to-amplitude DF: sum-and-difference processing class.

Referring to Figure 2.5, the phase delay, τ, across the baseline can be shown to be given by

$$\tau = (2\pi d/\lambda) \sin\phi \sin\theta \qquad (5.30)$$

The magnitude of the vector difference between τ and a unit vector, representing the signal at antenna 1, is the difference output Δ given by

$$\Delta = 2|\sin[(\pi d/\lambda) \sin\phi \sin\theta]| \qquad (5.31)$$

with E_1 and E_2 equal to unity. Equation (5.31) is the difference output depicted in Figure 2.6 for $\theta = 90°$, and it is the basic output for the single-baseline Adcock antenna using omnidirectional, vertical E-field elements. The output for a baseline orthogonal to the one for Equation (5.31) is given by

$$\Delta = 2|\sin[(\pi d/\lambda) \cos\phi \sin\theta]| \qquad (5.32)$$

The Δ response patterns defined by Eqs. (5.31) and (5.32) are substantially sinusoidal if $d/\lambda \ll 1$. Slight deviations from the desired $\sin\phi$ and $\cos\phi$ responses are not significant for the direct-amplitude null sensing techniques shown in Figure 5.19. The null locations are unaffected by the d/λ ratio until the ratio exceeds about unity. For example, for a rotatable Adcock DF system, the baseline distance may be about λ before the directional pattern becomes unusable. If d exceeds 0.5λ, the figure-eight pattern develops a dimple at mid-lobe, and the dimple becomes a null at $d = \lambda$ [30]. However, for nonrotating, amplitude-comparison, orthogonal baseline Adcock DF systems, the d/λ ratio should be maintained small relative to a wavelength. This is discussed further in Section 5.3.5.

The practical effect of an elevation angle θ of less than 90° is to reduce the effective d/λ ratio, which, in turn, enhances the deleterious effects of unbalance and spacing errors.

Equations (5.31) and (5.32) are based on the use of vertical E-field elements with azimuthal omnidirectional patterns. If H-field (loop) elements are used in a coaxial configuration, the composite azimuthal response pattern is modified by the directional patterns of the individual loop antennas.

The single-baseline, single-channel Adcock class provides several very significant improvements over the direct-amplitude Category I systems. The major strengths and limitations of the Adcock technique relative to the loop technique are as follows:

Strengths

- Reduced polarization error
- Improved sensing for some variants.

Limitations

- Quadrantal (spacing) errors
- Reduced portability
- Environmental susceptibility
- Accuracy decreases as θ decreases.

Another single-baseline, single-channel type is the probed line DF system shown in Figure 5.22 [31]. Two antennas are separated by a distance d and are directly connected through equal lengths of matched, lossless transmission lines to a probed line of length l. Parameters β_1 and β_2 are inherent phase shifts in the antenna elements and transmission lines.

The RF signal is a pure standing wave if the open-circuit antenna element output amplitudes are equal. The 180° phase shift at antenna 2 places the standing wave null at the center of the probed line when $\phi = 0°$. The null position is a function of ϕ but is independent of frequency if the probed line has a propagation constant equal to the free-space propagation constant. The probed line may be either a slotted line with a movable probe to sense the null position or multiple, fixed probes to sample the standing wave continuously and compute the null position. At least three fixed probes are required.

In Figure 5.22, the open-circuit voltages at the antenna terminals are given by

$$V_1 = A_1 \sin(\omega t - \zeta_1) \tag{5.33}$$

and

$$V_2 = -A_2 \sin(\omega t - kd \sin\phi - \zeta_2) \tag{5.34}$$

where A_1 and A_2 are the voltage amplitudes; k is the free-space propagation constant, $2\pi/\lambda$; and ζ_1 and ζ_2 are phase shifts inherent in the RF circuitry. Following Robertson [31], when $A_1 = A_2$, $\beta_1 = \beta_2$, and $\zeta_1 = \zeta_2$, the RF voltage on the probed line may be calculated from

$$\begin{aligned} V(z,t) = &(A_1/2) \sin[\omega t - k(-z + l/2) - \beta_1] \\ &- (A_2/2) \sin[\omega t - kd \sin\phi - k(z + l/2) - \beta_2] \end{aligned} \tag{5.35}$$

Using the trigonometric identity

$$\sin\alpha - \sin\beta = 2 \cos0.5(\alpha + \beta) \sin0.5(\alpha - \beta) \tag{5.36}$$

in Eq. (5.35) provides

$$V(z,t) = A_1 \cos[\omega t - kl/2 - (kd/2) \sin\phi] \sin[kz + (kd/2) \sin\phi] \tag{5.37}$$

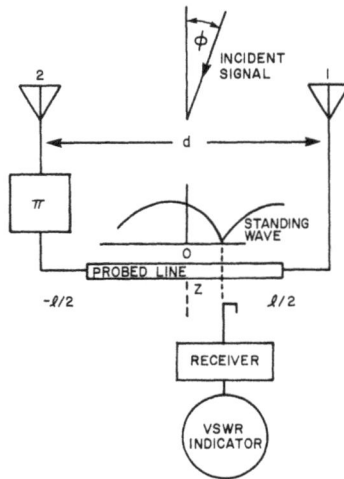

Figure 5.22 Probed line DF technique.

which is of the form $V(z,t) = G(t)F(z)$. Equation (5.37) shows that the voltage null locations are expressed by an amplitude function rather than an RF phase, i.e., phase-to-amplitude conversion occurs.

The null locations are found from the relationship

$$kz + (kd/2) \sin\phi = n\pi \qquad (5.38)$$

With $n \equiv 0$ in Eq. (5.38),

$$\sin\phi = -2z/d \qquad (5.39)$$

or

$$\phi = \arcsin(-2z/d) \qquad (5.40)$$

The null location, z, is a function of the angle-of-arrival, ϕ, and the physical antenna spacing, d. Hence, the null location is independent of frequency, and broadband operation is possible—if amplitude and phase matching can be maintained.

The effective element spacing is reduced by $\sin\theta$; therefore, performance on sky wave signals with $\theta < 90°$ is degraded.

Broadband operation and inherent simplicity are major attributes of the probed line DF technique; however, slow bearing acquisition, arrival angle ambiguity, and low sensitivity are major limitations. Sensitivity is reduced by the requirement for loose probe coupling to the probed line. The probed line DF system is most useful in the VHF and UHF bands for homing on cooperative transmissions of relatively long duration.

The single-baseline Adcock antenna may be electronically scanned to increase bearing acquisition time and eliminate the need for physical rotation of the array to the null. The null position of an Adcock antenna may be shifted by introducing a relative phase, or time delay, between the two antenna elements. Figure 5.23 plots the null angle, relative to the perpendicular to the baseline, *versus* the normalized relative time delay. The normalization factor, T_o, is the baseline distance, d, expressed in nanoseconds. (There are 0.9836 feet per nanosecond.) When $t/T_o = 0$, the response pattern is a figure-eight; when $t/T_o = 1$, the response pattern is a cardioid. The null has been shifted 90°. The basic principle is illustrated by Figure 5.24, which shows an Adcock antenna with a phase shift, τ, at the output of one element. When

$$\tau = kd \sin\phi \tag{5.41}$$

where $k = 2\pi/\lambda$, the output Δ of the differencing circuit is zero, and a null occurs. The null angle is

$$\phi = \arcsin(\tau/kd) \tag{5.42}$$

In the time domain, T_o is the time required for a signal arriving along the baseline axis $\phi \equiv 90°$ to traverse the aperture spacing d. Then

$$t_i = T_o \sin\phi \tag{5.43}$$

is the time required for a signal incident at an angle ϕ to traverse the aperture. If t_d is the time delay corresponding to a phase delay τ, a null occurs when $t_i = t_d$. Then

$$\phi = \arcsin[t_d/T_o] \tag{5.44}$$

Figure 5.25 depicts the functional diagram of the phase scanning DF technique [32]. The functions τ_v are due to voltage-variable delay lines. A null occurs when

$$T_o \sin\phi = T_o v(t) \tag{5.45}$$

because the difference in time delay between the two halves of the aperture is given by

$$\Delta t = [-(T_o/2) \sin\phi + (T_o/2)v(t)] - [(T_o/2) \sin\phi - (T_o/2)v(t)] \tag{5.46}$$

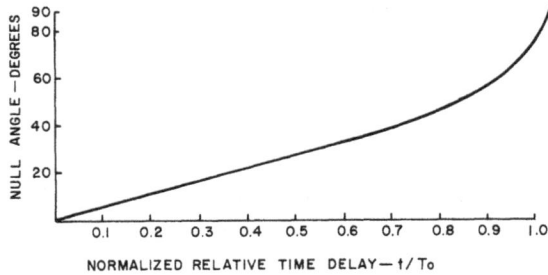

Figure 5.23 Probed line null angle *versus* normalized relative time delay.

In Eq. (5.46), when $v(t) = \sin\phi$, Δt is zero, and a null occurs. The angle ϕ is determined by correlating null occurrence with the time delay control voltage $v(t)$. When the receiver output is applied to the vertical deflection plates of a CRT and the phase scanning function $v(t)$ is applied to the horizontal plates, a null will be displayed at a CRT x-axis location corresponding to the azimuth angle ϕ. The displacement of the null is proportional to the sine of the relative bearing angle, and the null location is approximately linear to $\phi = \pm 45°$. CRT angular resolution is nearly constant to about $\phi = \pm 60°$, but degrades to zero at $\phi = \pm 90°$.

The receiver IF bandwidth must be three to four times the maximum scanning rate. At HF, a typical scanning rate is 100 Hz; at VHF, it is 1 kHz.

Bearing accuracy degrades as the elevation angle decreases from 90°. The accuracy degradation factor is $\sin\theta$.

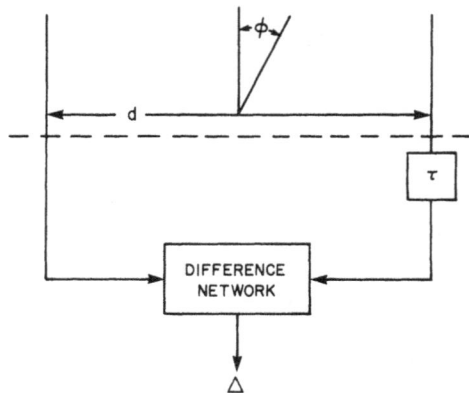

Figure 5.24 Adcock antenna with phase delay.

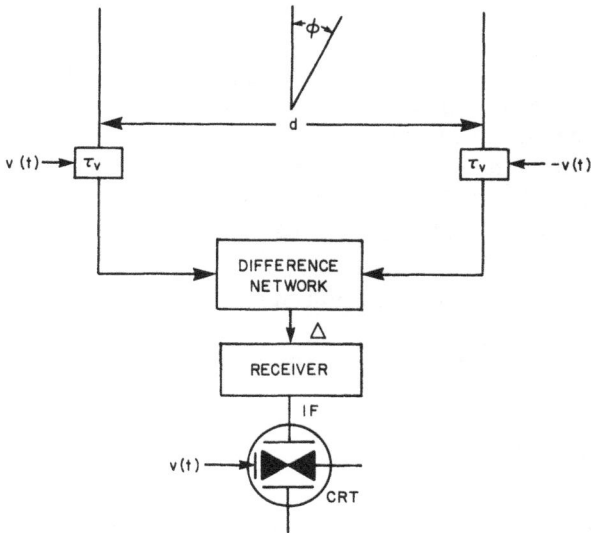

Figure 5.25 Phase scanning DF technique.

A sense ambiguity exists fore and aft of the perpendicular to the baseline. Sense may be acquired by physically rotating the baseline a slight amount. A forward relative bearing moves toward the center of the display; an aft bearing moves away from the center of the display.

The critical component in the phase scanning DF technique is the voltage-variable delay device that is described in Reference [31].

5.3.4 Dual-Baseline Antenna–Single-Channel Receiver

Figure 5.20 delineates the dual-baseline antenna–single-channel receiver class of phase differential-to-amplitude DF systems. The delineation is identical to that for orthogonal, crossed loops (Figure 5.2). The crossed loops may be replaced by crossed, orthogonal Adcock arrays to produce phase differential data from dual baselines. A crossed-baseline configuration and azimuthal response are shown in Figure 5.17, bottom. This class was developed to alleviate crossed-loop polarization errors while maintaining relatively rapid bearing acquisition. Another objective was to extend the basic dual-element antenna–single-channel receiver DF technique into the VHF band where crossed-loop implementation is difficult and relatively ineffective.

Type A: Electromechanical Scanning

Radio goniometer scanning is used; however, the dual-baseline disposition of the antenna elements creates spacing or octantal errors. Equations (5.31) and (5.32) are expressions for the phase differentials across orthogonal baselines. As d/λ increases from a small value, such as 0.1, the antenna patterns deviate from the desired $\sin\phi$ and $\cos\phi$ responses. These deviations create octantal errors when used with a goniometer. The goniometer angle, ϕ_g, at a bearing null is determined from

$$\tan\phi_g = [\sin(\pi d \sin\phi \sin\theta/\lambda)]/[\sin(\pi d \cos\phi \sin\theta/\lambda)] \tag{5.47}$$

The indicated bearing angle, ϕ_g, is not equal to the actual signal bearing angle, ϕ. The error $(\phi_g - \phi)$ is approximated by

$$(\phi_g - \phi) \approx 2.4°(\pi d \sin\theta/\lambda)^2 \sin 4\phi \tag{5.48}$$

Equation (5.48), which is a good approximation to about $d = 0.35\lambda$, shows that octantal errors are zero along the axis of the two baselines at azimuth angles of $n\pi/2$, where $n = 0, 1, 2, 3$, and 4 and at quadrant midpoints at angles of $n\pi/4$, where $n = 1, 3, 5$, and 7. Octantal errors are maximum at eight angles corresponding to $n\pi/8$, where n is an odd number between 0 and 15. Error maxima occur every odd multiple of 22.5° and alternate in sign. As a function of elevation angle, θ, octantal errors are maximum at $\theta = 90°$ corresponding to surface wave conditions.

Under maximum error conditions,

$$|\phi_g - \phi| \approx |(2.4°)(\pi d/\lambda)^2|, \quad \text{degrees} \tag{5.49}$$

or

$$|\phi_g - \phi| \approx |(23.7)(d/\lambda)^2|, \quad \text{degrees} \tag{5.50}$$

Figure 5.26 is a plot of Eq. (5.50), which shows that substantial bearing error occurs when the spacing factor d/λ exceeds about 0.15. Octantal error is contained by maintaining d/λ small (e.g., 0.15) at the maximum operating frequency. The small spacing reduces sensitivity. At a frequency where $d = \lambda/6$, an Adcock array using two disposed vertical dipoles has a gain approximately equivalent to one dipole. Below the $\lambda/6$ frequency, the gain decreases at a rate of 6 dB per octave of frequency, resulting in reduced sensitivity. Obviously, a design trade-off between octantal error and sensitivity is required for any specific application.

Knowing θ and λ, octantal error can be computed. If θ is known to be large, e.g., $> 80°$, octantal errors may be calculated assuming $\theta = 90°$ and confidently used for bearing error correction purposes. If θ is unknown or small, e.g., $< 80°$,

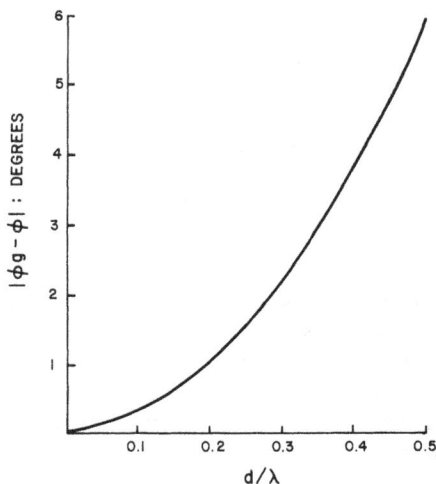

Figure 5.26 Octantal bearing errors *versus* baseline spacing.

corrections should not be applied. Octantal error correction is especially effective for applications operating over a narrow RF band on surface wave signals. For example, octantal error correction is effectively used in marine band VHF RDF systems in the 156- to 162-MHz band.

The operating RF bandwidth of the four-element dual-baseline Adcock DF antenna is limited by decreasing sensitivity at the lowest operating frequency and octantal spacing error at the highest operating frequency. For the four-element array, the operating frequency band is about 3 to 1, which is inadequate for many applications such as HF operation from 3 to 30 MHz—a 10-to-1 range. Error reduction and bandwidth extension have been obtained by use of the eight-element Adcock array [33]. These arrays consist of four pairs of antenna elements symmetrically disposed on the circumference of a circle. (Both E-field and H-field elements may be used.) Adjacent pairs are interconnected in parallel to form a single Adcock element, which is then connected out of phase with a diametrically opposite parallel pair. Orthogonal, dual-baselines are formed and scanned with a goniometer. Investigations [34–37] have developed the error-reduction and bandwidth capabilities of the eight-element Adcock antenna.

Pertinent strengths and limitations of the dual-baseline, single-receiver channel, electromechanically scanned class are as follows:

Strengths

- Greatly reduced polarization error.

Limitations

- Octantal errors
- Quadrantal errors
- Limited RF operating bandwidth.

Type B: Electronic Scanning

All of the electronic scanning techniques discussed in this section may be used with orthogonal, crossed-baseline Adcock arrays. The crossed loops are replaced by the Adcock array. Figure 5.27 shows an electronic analog scanning technique [38] for four elements disposed on two baselines oriented north–south and east–west. A single-channel receiver is used. Each antenna element feeds a balanced modulator that is electronically scanned. The RF signals from the antenna elements are balance modulated by time-quadrature scanning signals. Assuming a surface-wave incident signal, the RF output signals from the antenna elements are given by [39]

$$E_n = E_i h \sin(\omega_c t + \beta r \cos\phi) \tag{5.51}$$

$$E_e = E_i h \sin(\omega_c t + \beta r \sin\phi) \tag{5.52}$$

$$E_s = E_i h \sin(\omega_c t - \beta r \cos\phi) \tag{5.53}$$

$$E_w = E_i h \sin(\omega_c t - \beta r \sin\phi) \tag{5.54}$$

where

E_i = the incident signal amplitude;
h = the effective height of the antenna elements;
$\beta = 2\pi/\lambda$;
ω_c = the RF carrier frequency in radians;
r = the distance to each element from the center of the array; $2r = d$ is the baseline spacing;
ϕ = the azimuth angle of the signal measured clockwise from the north element.

The balanced modulator outputs are equal-amplitude, carrier-suppressed signals modulated by the analog scanning signals. These outputs are given by the following expressions:

$$E'_n = E_i K \sin(\omega_c t + \beta r \cos\phi) \cos\omega_s t \tag{5.55}$$

$$E'_e = E_i K \sin(\omega_c t + \beta r \sin\phi) \sin\omega_s t \tag{5.56}$$

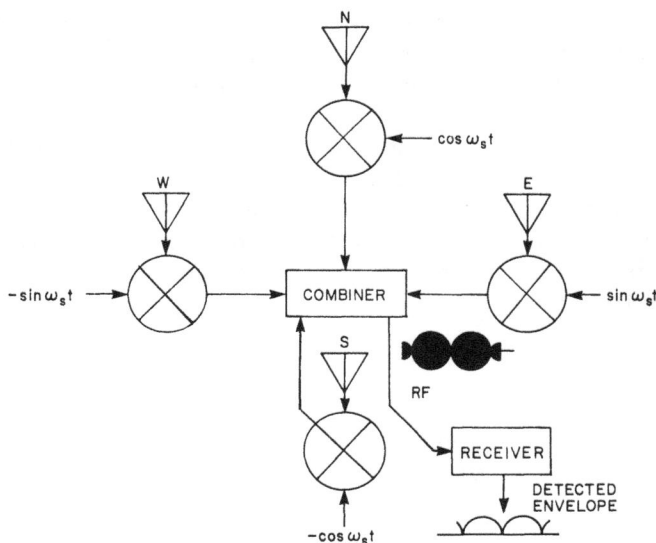

Figure 5.27 Electronic analog scanning DF technique.

$$E'_s = -E_iK \sin(\omega_c t - \beta r \cos\phi) \cos\omega_s t \tag{5.57}$$

$$E'_w = -E_iK \sin(\omega_c t - \beta r \sin\phi) \sin\omega_s t \tag{5.58}$$

where K is h times the balanced modulator gain and ω_s is the scanning frequency in radians. Linear summation of Eqs. (5.55) through (5.58) yields a resultant E_r given by

$$E_r = 2E_iK[\sin(\beta r \cos\phi) \cos\omega_s t + \sin(\beta r \sin\phi) \sin\omega_s t] \cos\omega_c t \tag{5.59}$$

For $\beta r \ll 1$,

$$E_r \approx 2\beta r E_iK \cos(\omega_s t - \phi) \cos\omega_c t \tag{5.60}$$

Equation (5.60) shows that the input to the receiver is a carrier-suppressed, modulated signal with the bearing information contained in the envelope phase. Conventional envelope detection (or synchronous detection if an additional receiver is used) of the IF output converts the envelope phase to a more usable form for synchronous demodulation and, hence, bearing angle derivation. Bearing ambiguity is resolved by using a nondirectional signal obtained from a summation of the outputs of all elements.

Major design considerations are intermodulation and crossmodulation effects in the RF-balanced modulators, the requirement for matched modulation, and the preservation of the envelope phase through the receiver from RF input to the IF output.

A variant of the basic electronic scanning technique is shown in Figure 5.28 The phase differentials across the baselines are converted to amplitudes representing $\sin\phi$ and $\cos\phi$. The amplitudes are scan modulated by carrier-suppressed signals at two different frequencies, which typically lie in the audio band (for example, 5 and 6 kHz). The scan frequencies are eliminated from the desired intelligence by sharp rejection-band filters.

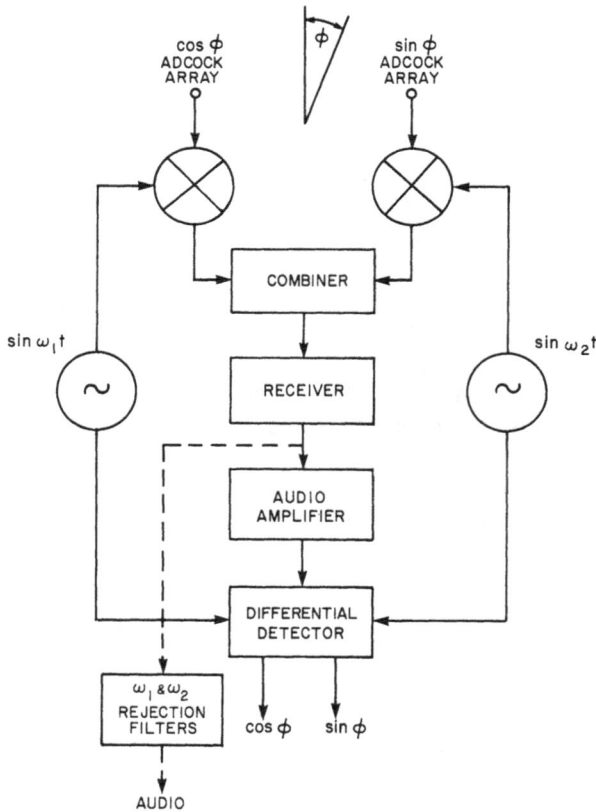

Figure 5.28 Adcock array electronic analog scanning DF technique.

Figure 5.29 shows another form of electronic scanning, which is applied before the differential phase function has been formed. Offset local oscillators (LOs) are used with one LO set high by a frequency ω_s and one LO set low by a frequency ω_s. The frequency ω_s is the scanning frequency, which is selected to reduce interference to the signal intelligence. Combining the mixer outputs creates a suppressed carrier signal centered at ω_o with the modulation envelope phase containing the bearing data. Up-conversion is normally used, and tuning is performed by changing the LO frequency. The receiver is fixed tuned. This technique is inherently narrowband and requires precise RF channel and LO matching.

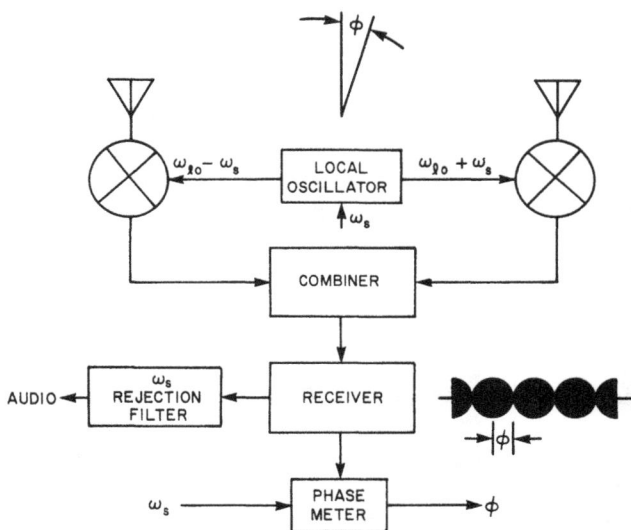

Figure 5.29 Offset local oscillator DF scanning technique.

Another design consideration is the effect of the scan frequency ω_s on signal modulation components and intelligence. If ω_s were at frequencies greater than the signal modulation sidebands, interference could be reduced; however, excessive scan modulation envelope phase shift may occur in the receiver, creating bearing error. The scan modulation is normally set at the high end of the signal modulation spectrum, and narrowband filters are used to remove the scan frequency ω_s with minimal effects on the signal intelligence. This technique operates best for specialized applications using cooperative emitters with good frequency stability. This technique was used on the Naval Research Laboratory minitrack satellite network for the Vanguard and Mercury programs [40]. It has also been used for wildlife tracking by the U.S. Department of Interior.

The pertinent strengths and limitations of dual-baseline, single-receiver channel, electronically scanned DF systems are as follows:

Strengths

- Rapid bearing acquisition
- Reduced polarization error
- No matched channel receivers needed.

Limitations

- Octantal spacing errors (mechanical goniometers are used)
- Quadrantal errors
- Precision RF matching and tracking required (or near real-time calibration)
- RF intermodulation and crossmodulation in RF scanning circuits
- Reduced accuracy on sky wave signals
- Limited RF operating range
- Possible interference to signal modulation.

5.3.5   Single- and Dual-Baseline Antenna–Dual-Channel Receiver

Figure 5.21 defines the structure of the single- and dual-baseline antenna–dual-channel receiver class of small-aperture DF systems. This class uses RF sum (Σ) and difference (Δ) processing prior to the dual-channel receiver. The two dominant types are an instantaneous DF technique and a mechanical scanning technique. The basic operating principles are described below.

Figure 5.30 shows the basic elements of the sum-and-difference technique. The vertical antenna element spacing, d, is less than $\lambda/2$. The incident signal is $E_i \exp(j\omega_c t)$, and the RF antenna outputs are given by

$$E_1 = E_1' \exp[(jkd \sin\phi \sin\theta)/2] \exp(j\psi_1) \qquad (5.61)$$

and

$$E_2 = E_2' \exp[(-jkd \sin\phi \sin\theta)/2] \exp(j\psi_2) \qquad (5.62)$$

where the RF carrier term $\exp(j\omega_c t)$ is assumed, and the RF phase across is referenced to the baseline midpoint. In Equations (5.61) and (5.62), k is the free-space

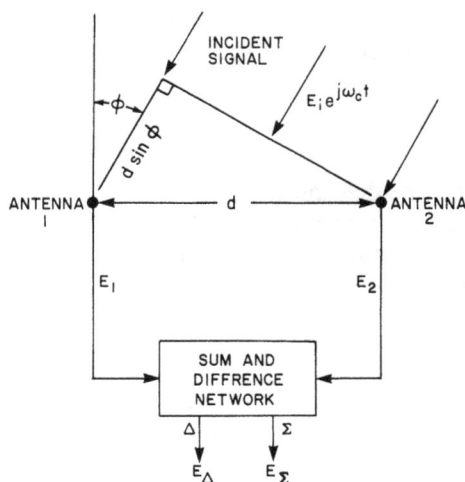

Figure 5.30 Basic elements of sum-and-difference DF techniques.

propagation constant, $2\pi/\lambda$; θ is the elevation angle of arrival; and ψ_1, ψ_2 are the phase shifts created by factors such as transmission lines, coupling networks, complex antenna impedances, and environmental coupling.

Assuming $E_1' = E_2'$ and $\psi_1 = \psi_2$, the difference output, E_Δ, is given by

$$E_\Delta = 2E_1' \sin[(\pi d \sin\phi \sin\theta)/\lambda] \tag{5.63}$$

and the sum output, E_Σ, is

$$E_\Sigma = 2E_1' \cos[(\pi d \sin\phi \sin\theta)/\lambda] \tag{5.64}$$

The E_Δ/E_Σ ratio is

$$E_\Delta/E_\Sigma = \tan[\pi d \sin\phi \sin\theta)/\lambda] \tag{5.65}$$

and the azimuth angle of arrival may be computed from

$$\phi = \arcsin\{(\lambda/\pi d)(\csc\theta)[\arctan(E_\Delta/E_\Sigma)]\} \tag{5.66}$$

If the elevation angle of arrival is large (e.g., $\theta > 60°$), ϕ is approximated by

$$\phi \approx \arcsin[(\lambda/\pi d) \arctan(E_\Delta/E_\Sigma)] \tag{5.67}$$

The angle of arrival ϕ can be calculated because λ and d are known and (E_Δ/E_Σ) can be measured and converted to the ratio (V_Δ/V_Σ).

Type A: Instantaneous DF

The angle of arrival ϕ can be determined instantaneously without scanning. Figure 5.31 depicts a functional approach using both an angle-of-arrival computer and a CRT display of the parameter δ, which is an analog of ϕ. For $\phi \ll \pi/2$ and $\theta \approx 90°$ $\delta = K\phi$, where $K = \pi d/\lambda$. Hence,

$$\phi = \delta\lambda/\pi d \qquad (5.68)$$

Figure 5.32 presents a diagram of an RF processing circuit that provides in-phase and quadrature sum-and-difference RF outputs. After dual-channel down-conversion and processing, IF representations of the 180° hybrid outputs may be combined to obtain the following expressions:

$$R_1 = \cos[(2\pi d/\lambda)(\sin\phi)] = (|V_\Sigma|^2 - |V_\Delta|^2)/(|V_\Sigma|^2 + |V_\Delta|^2) \qquad (5.69)$$

or

$$\phi = \arcsin[(\lambda/2\pi d)\arccos(R_1)] \qquad (5.70)$$

where V_i is the IF equivalent of E_i, and an elevation angle of $\theta = 90°$ is assumed Similarly,

$$R_2 = \sin[(2\pi d/\lambda)(\sin\phi)] = (|V_{\Sigma j}|^2 - |V_{\Delta j}|^2)/(|V_{\Sigma j}|^2 + |V_{\Delta j}|^2) \qquad (5.71)$$

or

$$\phi = \arcsin[(\lambda/2\pi d)\arcsin(R_2)] \qquad (5.72)$$

where V_{ij} is the IF equivalent of E_{ij}, and an elevation angle of $\theta = 90°$ is assumed.

Other configurations of quadrature and 90° hybrids may be used to obtain an instantaneous measure of the space phase difference across the baseline of two disposed antenna elements feeding separate channels of a matched dual-channel receiver. Figure 5.33 illustrates two direct measurement techniques that provide measures of the space phase difference, δ. In Figure 5.33(a), an IF hybrid phases and combines the two IF signals producing two other IF signals, V_a and V_b. If a 180° hybrid is used, the space phase difference, δ, is given by

$$\delta = \text{arc } \cos[(|V_b|^2 - |V_a|^2)/(2|V_1||V_2|)] \qquad (5.73)$$

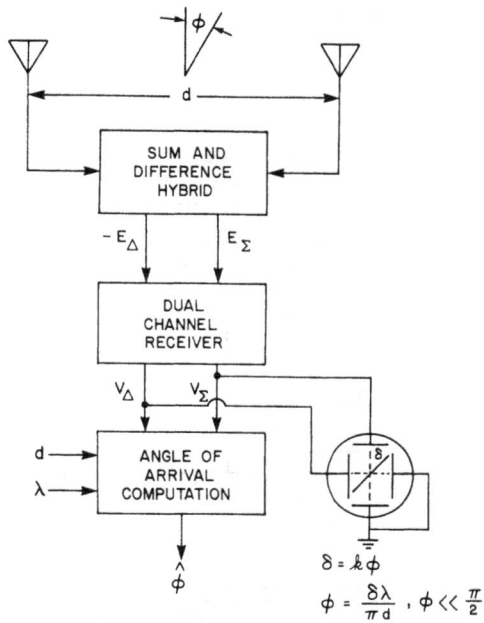

Figure 5.31 Instantaneous sum-and-difference DF technique.

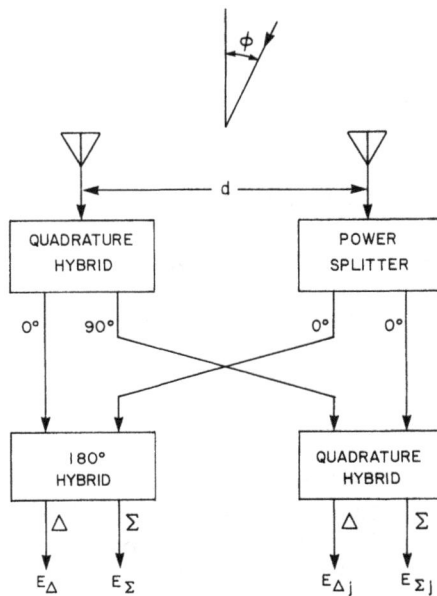

Figure 5.32 Sum-and-difference RF processing circuit.

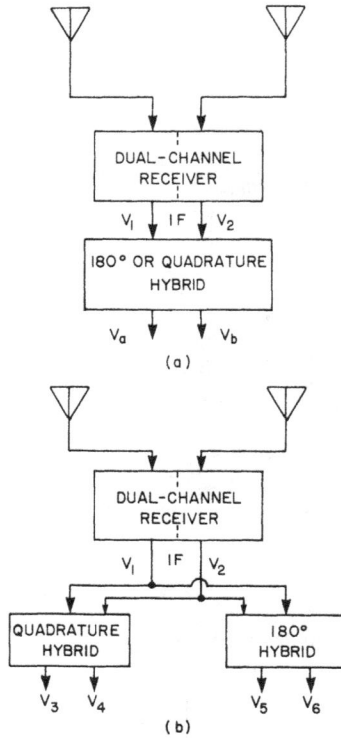

Figure 5.33 IF processing circuits.

If a quadrature hybrid is used, δ is given by

$$\delta = \arcsin[(|V_a|^2 - |V_b|^2)/(2|V_1||V_2|)] \tag{5.74}$$

If both quadrature and 180° hybrids are used, as shown in Figure 5.33(b), δ is computed from

$$\delta = \arctan[(|V_3|^2 - |V_4|^2)/(|V_5|^2 - |V_6|^2)] \tag{5.75}$$

The techniques using passive RF and IF processing by hybrid configurations are theoretically attractive and operationally appealing. Indeed, at microwave frequencies, passive RF processing via hybrids is widely used; however, for small-aperture DF systems operating in the MF and HF bands, practical design considerations limit effective application.

One major design consideration is the requirement for either high integrity amplitude and phase balance or frequent measurement and calibration of unbalance effects by RF front-end signal insertion. (This is discussed further in Section 5.4.6.) Another major design consideration is the effects of the relatively high insertion loss of the hybrids. Multihybrid, prereceiver processing can have as much as a 10-dB total RF loss including both the coupling loss and insertion loss. A result is significant sensitivity reduction. Systems of this type operate best on cooperative transmissions with relatively large signal levels.

Pertinent strengths and limitations of the single-baseline, dual-receiver channel class are as follows

Strengths

- Instantaneous bearing acquisition
- No electromechanical or electronic scanning required
- Reduced polarization errors
- Low power requirements.

Limitations

- Requires either RF processor and receiver channel matching or near-real time calibration.
- Limited azimuthal FOV
- Poor modulation tolerance
- High RF losses
- Performance degrades as elevation angle decreases
- Requires knowledge of elevation angle of arrival
- Contains angle-of-arrival ambiguity.

Type A: Dual-Baseline–Dual-Channel Receiver

A major limitation of dual-element antenna–single-baseline, phase-differential DF techniques is that the angle of arrival cannot be determined until the element spacing factor and elevation angle are known and applied as correction factors. Further, the fore-aft signal angle-of-arrival ambiguity must be resolved by the use of a separate sense antenna. Also, the bearing error varies as a function of ϕ. The most accurate FOV is limited to about $\pm 60°$ about the perpendicular to the baseline. (The required FOV for any application is a function of the maximum allowable bearing error. For example, for homing systems, the maximum allowable bearing error may be much larger than those for stand-off DF systems requiring relatively precise position location.) These major limitations may be alleviated by (1) using two orthogonal apertures, requiring a minimum of three elements, (2) sequentially measuring the space phase differential across the two apertures, and (3) calculating the azimuth angle of arrival ϕ from the expression

$$\phi = \arctan(\delta_1/\delta_2) \qquad (5.76)$$

where δ_1 and δ_2 are the space phases across apertures 1 and 2, respectively. The elevation angle of arrival may be obtained using the expression

$$\theta = \arcsin[(\sqrt{\delta_1^2 + \delta_2^2})/(\beta d)] \qquad (5.77)$$

where $\beta = 2\pi/\lambda$ and d is the baseline spacing, which is assumed to be the same for each aperture.

Pertinent strengths and limitations of the dual-baseline, dual-channel receiver class are as follows:

Strengths

- Instantaneous bearing acquisition
- No electromechanical or electronic scanning required
- Reduced polarization errors
- Low power requirements
- Omnidirectional FOV
- Elevation angle-of-arrival measure
- No angle-of-arrival ambiguity.

Limitations

- Requires either phase and amplitude match or near real-time calibration of mismatch
- Poor modulation tolerance
- High RF losses
- Performance degrades as elevation angle decreases.

Type B: Mechanical Scanning

Figure 5.34 diagrams a sum-and-difference technique using mechanical rotation of a two-element array [41]. The rotation is continuous about a vertical axis, and the rotation rate is ω_s in radians per second. The sum (Σ) and difference (Δ) outputs of the rotating array are down-converted to IF signals and synchronously detected

[42]. The synchronous detector output is low-pass filtered to obtain a baseband sinusoid. The zero-crossing of the sinusoid contains the bearing angle ϕ information. In Figure 5.34, K is a function of d, λ, θ, and incident signal polarization angle; K is also a function of ϕ if spaced-loop antenna elements are used. The azimuth angle of arrival ϕ is derived by synchronously detecting $\cos(\omega_s t + \phi)$ with an analog of the rotation rate—such as $\cos(\omega_s t)$.

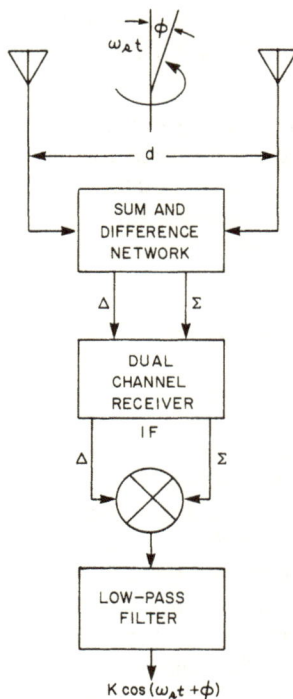

Figure 5.34 Sum-and-difference technique using mechanical rotation.

The mechanically rotated sum-and-difference DF systems has several significant unique strengths as follows:

1. When vertical dipoles are used, increased rejection of low elevation angle sky waves occurs as the array multiplication factor is given by $\sin^3\theta$. As θ decreases, the array response decreases as a cube of θ. Near zenith, the response is essentially zero. This is a major advantage when operating in a combined ground wave–sky wave situation, because enhanced rejection of the sky wave component is possible.

2. Synchronous detection improves sensitivity; the theoretical improvement is 3 dB. In practice, the improvement may be approximately 2 dB [42].
3. Synchronous detection improves the sharpness and readability of a CRT bearing display.
4. Bearing sense is inherent in the synchronous detector output.
5. Matched channels are not required. The only requirement on phase tracking is that the undesired phase shift between channels be maintained within about $\pm 45°$.
6. Bearing accuracy is maintained over a 360° FOV because accuracy is independent of the bearing angle ϕ. All DF information is acquired on the perpendicular to the baseline.

Limitations are significant. The two major limitations are (1) relatively high bearing acquisition time and (2) the power required for mechanical rotation. Other disadvantages are the low reliability of mechanical rotation and the associated maintenance problems. Further, mechanically rotated arrays are difficult to conceal from acoustic, optical, and radar sensors.

5.3.6 Dual-Baseline with Sense Antenna–Triple-Channel Receiver (Adcock–Watson-Watt)

The Adcock–Watson-Watt DF system diagrammed in Figure 5.35 provides a virtually instantaneous DF capability with inherent angle-of-arrival ambiguity resolution. The system uses the Watson-Watt DF technique described in Section 5.2.4 combined with orthogonal baselines using Adcock antennas and a phase-center sense antenna. The phase differentials across the two baselines are converted to amplitude responses as illustrated by Figure 5.36. The north-south, E_{NS}, east-west, E_{EW}, and sense, E_S, amplitudes are applied to a triple-channel receiver with closely matched phase and gain characteristics. After conversion to IF, the IF signals V_{EW} and V_{NS} are applied to the xy CRT plates, respectively. In general, an ellipse is formed on the CRT display with the major axis laying along the azimuth bearing. Under ideal conditions, V_{EW} and V_{NS} are 180° out of phase, and the ellipse becomes a line oriented on the bearing. Sense information from V_S is applied to the CRT z-axis modulation to unblank the true bearing direction and blank out the ambiguous half of the ellipse major axis. Polarization rotation and multipath propagation broaden the elliptical display and create bearing inaccuracy.

The IF outputs shown in Figure 5.36 may also be applied to digital and analog processing circuitry to obtain a bearing measure. A CRT is not necessary; however, it is highly desirable due to the virtually instantaneous response and the indication of signal quality inherent in the ellipse display.

The response time is limited only by the response (settling) time of the receiver channels, display processor, and any measurement time interval.

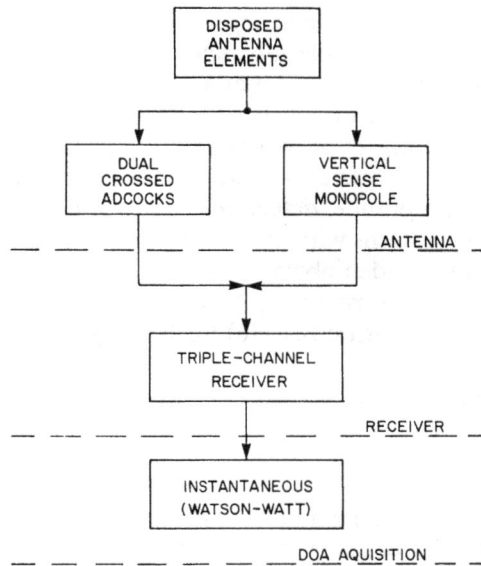

Figure 5.35 Phase differential-to-amplitude DF: Adcock–Watson-Watt class.

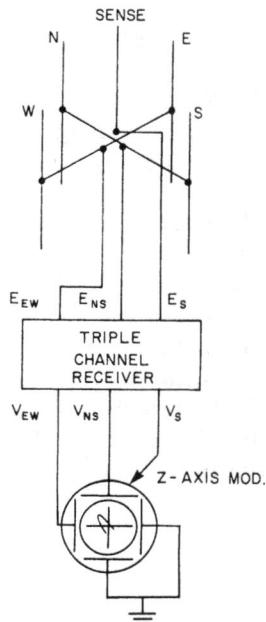

Figure 5.36 Triple-channel Adcock–Watson-Watt system.

The major bearing error source is amplitude unbalance between the NS and EW receiver channels and amplitude mistracking between the orthogonal, figure-eight NS and EW response patterns. Antenna and receiver amplitude unbalance effects are additive, and the effect on bearing error has been investigated in considerable detail [43–46]. Bearing error in degrees is a linear function of amplitude unbalance in decibels. The exact ratio is a function of antenna pattern characteristics and the type of amplitude detection and post-detection processing. For orthogonal sine antenna patterns with an IF CRT display, the ratio is cited as being 1° error for a 0.35-dB amplitude unbalance or 2.9° error for a 1-dB unbalance. For quadratic detection, the cited error is 3.2° for a 1-dB amplitude unbalance error [43]. A commonly used expression [44–46] for bearing error is

$$\Delta\phi = (\phi_{BW})(\Delta A)/24.1 \qquad (5.78)$$

where

$\Delta\phi$ = the peak bearing error in degrees;
ϕ_{BW} = the beamwidth of the patterns in degrees, which is 90° for an orthogonal sine pattern;
ΔA = amplitude mismatch or unbalance in decibels.

Equation (5.78) provides a ratio of 3.73° error per 1-dB amplitude unbalance. All error computation techniques indicate that very careful attention must be given to maintaining amplitude balance between the NS and EW amplitude responses.

The Adcock–Watson-Watt method is emerging as the most effective DF technique for spread-spectrum, frequency-hopping transmissions. The virtually instantaneous DF capability combined with efficient frequency-scanning receivers provides for the intercept and DF of frequency-hoppers. If both space and frequency must be scanned in time, the intercept and direction finding of typical HF and VHF frequency-hopping transmissions are very difficult.

The CRT display provides for bearing information on noncoherent, co-channel interference. When two signals at the same frequency occur at the same time, the Adcock–Watson-Watt CRT bearing display is a parallelogram with the two sides indicating the azimuths to the two co-channel signals. Fading and polarization rotation may compromise the integrity of the parallelogram.

The strengths and limitations of the Adcock–Watson-Watt DF technique are as follows:

Strengths

- Instantaneous angle-of-arrival information
- Inherent sense resolution

- Full azimuthal FOV
- Relatively tolerant of modulation
- CRT display of noncoherent, co-channel interference bearing
- Relatively simple implementation in theory.

Limitations

- Requires precise gain and phase match and track or frequency calibration
- Requires good siting to maintain antenna pattern integrity or thorough site calibration data and correction.

5.3.7 Instrumental and Observational Errors

Instrumental

Phase differential-to-amplitude DF techniques experience four major AOA error sources as follows:

1. Array phase unbalance;
2. Interelement scattering and coupling;
3. Octantal effects;
4. Quadrantal effects.

The first two error sources apply to all phase differential-to-amplitude DF techniques. Octantal errors occur only in techniques using tangent function derivation of bearing. Quadrantal errors occur in crossed-baseline systems using tangent function derivation of bearing. (Octantal and quadrantal errors were discussed in Section 5.3.4.)

Angle-of-arrival errors are introduced by phase unbalance in the antenna elements and circuitry preceding the sum-and-difference devices. Amplitude unbalance does not create an AOA error directly, as does phase unbalance, but amplitude unbalance does decrease the differential phase gradient around the difference null phase-transition point. Hence, amplitude unbalance does indirectly reduce accuracy.

The bearing error, ϕ_e, created by phase unbalance may be calculated using the expression for space phase delay

$$\delta = (2\pi d/\lambda) \sin\phi \sin\theta \tag{5.79}$$

and then expressing ϕ as a function of δ to obtain

$$\phi = \arcsin[(\delta\lambda)/(2\pi d \sin\theta)] \tag{5.80}$$

Errors in the bearing angle, ϕ, and phase differential, δ, may be related by the incremental expression

$$\Delta\phi = (\partial\phi/\partial\delta)\,\Delta\delta \qquad (5.81)$$

where the partial derivative is obtained by differentiating Eq. (5.80). The incremental quantities in Eq. (5.81) are assumed to be zero mean random variables; hence, the standard deviations or rms errors are related by

$$\sigma_\phi = (\sigma_\delta)/[(2\pi d/\lambda)\,\cos\phi\,\sin\theta] \qquad (5.82)$$

Equation (5.82) is often expressed as

$$\phi_\epsilon = (\delta_\epsilon)/[(2\pi d/\lambda)\,\cos\phi\,\sin\theta] \qquad (5.83)$$

where ϕ_ϵ is the bearing rms error in degrees (assumed to be a constant) and δ_ϵ is the phase rms unbalance or error in degrees (also assumed to be a constant). Equation (5.83) shows that the bearing error ϕ_ϵ increases as ϕ increases, i.e., as ϕ approaches a parallel path with the baseline. Signal arrival along the perpendicular to the baseline produces the smallest error because $\phi = 0°$ and $\cos\phi = 1$ along the perpendicular. As either d/λ decreases or θ decreases, the bearing errors increase for a fixed δ_ϵ value. For small d/λ apertures and typical phase unbalance values, the bearing error may be relatively large. For example, Figure 5.37 shows bearing error as a function of differential phase error for a signal arriving along the perpendicular to the baseline with $\theta = 90°$. A typical δ_ϵ value is 2°, which produces a 3.2° bearing error on a baseline of $d/\lambda = 0.1$. For very short baselines, such as $d/\lambda = 0.1$, δ_ϵ must be reduced to much less than 1° to maintain bearing errors of less than 2°, assuming surface waves arrive along the perpendicular to the baseline. The effect of elevation angle is depicted in Figure 5.38 for $d/\lambda = 0.1$ and a typical phase unbalance of 2°. Clearly, very precise phase balance is required for short-range ($\theta < 30°$) stand-off DF operation on sky wave transmissions. Homing DF applications can tolerate greater phase unbalance errors; however, high integrity phase balance aids faster homing.

Accurate measurements of phase unbalance by test signal insertion at RF is difficult because a significant amount of the phase unbalance usually occurs in the antenna elements, transmission lines, and couplers prior to a convenient and reliable test signal insertion point. The use of antenna element interchange and subsequent phase measurements can isolate phase unbalance occurring after the switching point. Test signal insertion and antenna element interchange have been successfully used to negate phase unbalance errors in small-aperture DF systems.

Mutual scattering and coupling between antenna elements can be a major source of phase-differential error in small-aperture DF systems. Mutual coupling

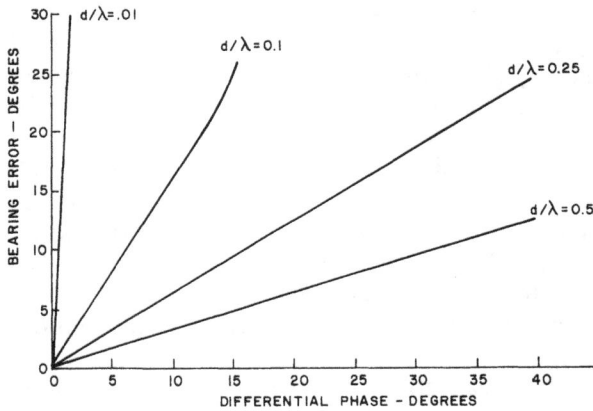

Figure 5.37 Differential phase error effects on bearing error.

and impedance between spaced antennas have been extensively treated [47–50], and are readily computed by moment method analysis techniques, which are discussed in the Appendix.

Antenna scattering errors in two-element and four-element disposed antenna, phase-differential DF systems have been analyzed by Harrison in References [47]

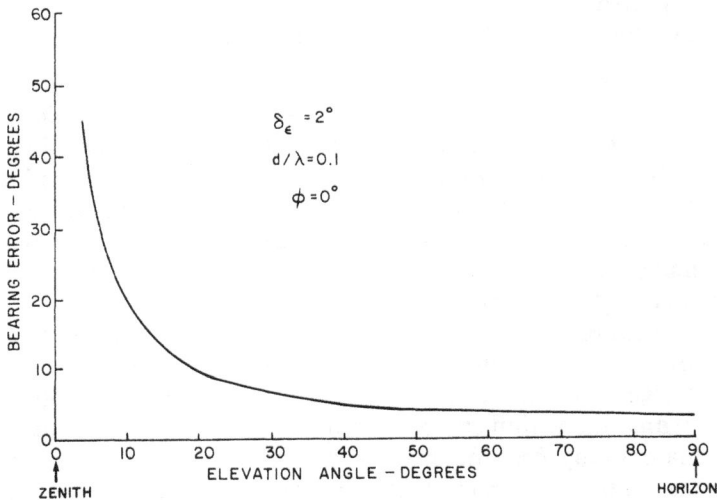

Figure 5.38 Elevation angle effects on bearing error.

and [48], respectively. Harrison's derivations are based on scattering effects. The electric field acting on any one DF antenna element in an array of elements is the vector sum of the direct signal from the RF source and the scattered signals maintained by the currents in the other DF elements. Thus, phase differentials may be in error for most signal angles of arrival. Harrison's derivations pertain to thin, vertical, free-space, resonant dipoles with load resistances of 25, 100, 400, and 1000 Ω. The two-element model consists of identical, parallel, nonstaggered antennas disposed on a baseline; the four-element model uses identical antennas oriented at the corners of a square. The incident signal is assumed to be vertically polarized parallel to the axis of the antennas. The azimuth angle of arrival is varied over one quadrant and the baseline spacing (diagonal spacing for the four elements) is varied from λ to 0.1λ.

Harrison's analysis leads to the following general conclusions:

1. No scattering errors exist along the perpendicular to the baseline or diagonal where $\phi = 0°$.
2. Maximum scattering errors exist when $\phi = \pi/2$ and the signal arrives along an array axis.
3. Scattering errors decrease as the antenna load impedance increases.
4. Scattering errors increase as the baseline spacing in wavelengths decreases. Relatively large increases occur as the spacing decreases below about 0.5λ.
5. Scattering errors exhibit minimum values at baseline spacings around 0.4λ. For large load impedance levels, the minimums are broad, ranging from about 0.35λ to 0.6λ.
6. Scattering errors for the four-element configuration are, in general, larger than those for the two-element configuration under similar conditions.

Figure 5.39 presents a plot of differential phase error *versus* antenna spacing in wavelengths for a two-element configuration. Signal arrival is parallel to the baseline with $\phi = 90°$. Load impedance levels are 100 to 1000 Ω. (The dotted lines on Figure 5.39 are data obtained from a model based on the equivalent lumped-constant network of two coupled antennas with a common load impedance [51].)

Figure 5.39 indicates that errors can be tens of degrees for low impedance levels and small antenna spacing. The smallest errors occur in the vicinity of 0.5λ for both low and high load impedance levels.

The analytical results of Harrison and others quantitatively verified the effectiveness of antenna interaction mitigation techniques that had been practiced by DF system designers for many decades. Early DF practitioners discovered that undesired antenna interaction can be alleviated by using nonresonant antenna elements and maintaining baseline spacing greater than about 0.3λ. Further, they found that the undesired effects of unused antenna elements in a multielement array can be greatly reduced by open-circuiting the unused elements. Also, increas-

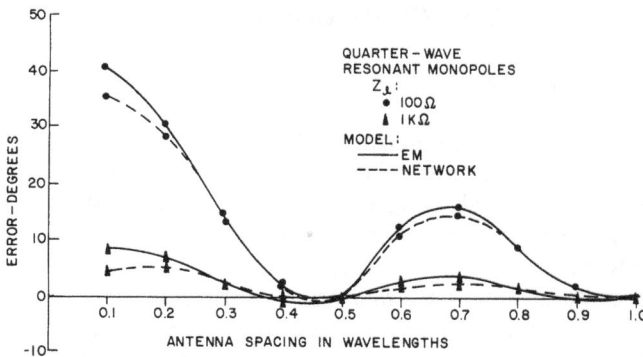

Figure 5.39 Antenna scattering and coupling errors *versus* baseline spacing.

ing element load impedance reduces interaction. (These practices are not effective if the open-circuited element becomes self-resonant.)

The penalties paid for antenna interaction reduction techniques are reduced sensitivity, because nonresonant elements are used, and limited RF operating bandwidth, because the allowable baseline spacing range is limited to about 0.3λ to 0.5λ—a 1.67 bandwidth.

Antenna scattering errors are systematic, deterministic, and explicit in calibration measurement data. Therefore, antenna scattering errors may be delineated in the direction-of-arrival error budget. Then compensation techniques are applied. The usual practice is to, first, attain a reasonable engineering compromise between antenna scattering error reduction and other system performance parameters, such as RF operating bandwidth and sensitivity, and, second, apply calibration correction data to remove the antenna scattering errors.

Quadrantal errors occur in crossed, dual-baseline systems using (1) either electromechanical or electronic scanning and (2) bearing derivation by amplitude-response ratio comparison based on tangent function algorithms. Also, physical misalignment and disorientation of the crossed baselines create quadrantal errors.

Observational

Direction-of-arrival read-out methods for phase-to-amplitude DF techniques involve aural-null and visual displays including CRTs, analog meters, and digital read-outs. Observational errors are very similar to those for amplitude DF techniques; however, the adverse effects of polarization rotation and signal modulation are considerably less.

5.4 CATEGORY III: PHASE COMPARISON

5.4.1 Functional Approach

Phase comparison DF systems determine DOA information by direct phase comparison of the subject signal received by separate, disposed antennas [52–57]. Figure 2.7 depicts the functional approach. Figure 5.40 shows the geometry for two disposed antenna elements with baseline spacing d and an incident signal at an angle ϕ relative to the boresight zy plane. The elevation angle is θ. The propagating wavefront is received at antenna 1 first and travels an additional distance ($d \sin\phi \sin\theta$) to antenna 2. During the time required to travel the distance between antennas 1 and 2, the phase of the signal received at antenna 1 changes by an amount δ given by

$$\delta = (2\pi d/\lambda) \sin\phi \sin\theta \qquad (5.84)$$

The azimuthal angle of arrival is obtained from

$$\phi = \arcsin[(\delta\lambda)/(2\pi d \sin\theta)] \qquad (5.85)$$

where δ is measured and λ, d, and θ are known. If the elevation angle θ is unknown, a value must be assumed if a single baseline is being used. However, if a dual-baseline system is used, both azimuth and elevation angles of arrival may be independently determined. The total phase delays across two baselines that are orthogonal in the azimuthal plane provide complete DOA information through the expressions

$$\phi = \arctan[\delta_1/\delta_2] \qquad (5.86)$$

for azimuth angle of arrival and

$$\theta = \arcsin[\sqrt{(\delta_1^2 + \delta_2^2)}(\lambda)/(2\pi d)] \qquad (5.87)$$

for elevation angle of arrival where

δ_1 = the phase delay across baseline 1;
δ_2 = the phase delay across baseline 2;
d = the baseline spacing, which is assumed the same for each baseline;
λ = the operating wavelength;

and the azimuth angle ϕ is referenced to baseline 2.

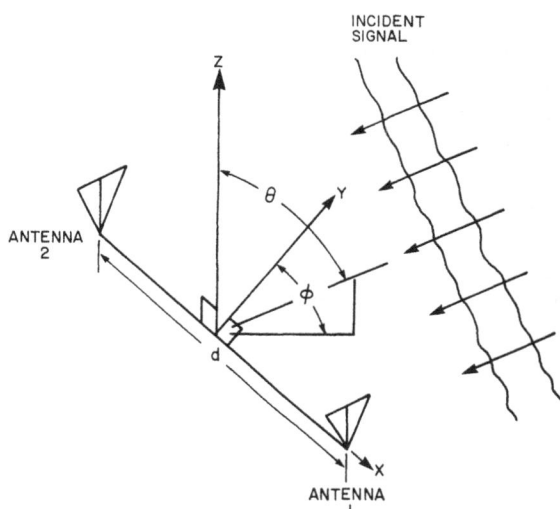

Figure 5.40 Geometry of phase comparison DF.

Direct phase comparison systems (interferometers) are often "long-baseline interferometers" in which the interelement spacings are large compared to wavelength. This increases phase measurement precision at the expense of direction-of-arrival ambiguity. To avoid ambiguities, the maximum phase shift δ must be within π radians for a maximum spacing of $d_{max} = \lambda/2$ [58]. Therefore, a phase comparison DF system with a baseline within $\lambda/2$ has no DOA ambiguities created by spacing. Long-baseline ambiguity resolution requires the use of amplitude measures obtained from interactions between the incident signal and the DF array. Small-aperture, short-baseline interferometers avoid the need for amplitude measurements at the expense of phase measurement precision and operating bandwidth.

DOA measurement precision is a function of $\Delta\delta/\Delta\phi$ in degrees per degree. As the baseline spacing increases, the function $\Delta\delta/\Delta\phi$ increases and phase delay can be measured with greater precision.

The RF operating range of a short-baseline interferometer is limited on the high frequency end by the $d = 0.5\lambda$ limit and on the low frequency end by the maximum tolerable phase errors, interelement scattering and coupling errors, minimum allowable SNR, and the availability of phase error compensation techniques. For a high SNR, robust processing, and efficient near real-time error compensation, the RF operating bandwidth may approach 5:1 corresponding to baseline spacings ranging from 0.1λ to 0.5λ. Multiple baselines are required to cover the decade-wide

HF band from 3 to 30 MHz. Important VHF regions, such as the 30- to 76-MHz and 56- to 162-MHz bands, may be covered by a single-baseline spacing.

In the microwave region, short-baseline interferometers are used with long-baseline interferometers to resolve ambiguities [55, 58]. Multiple short-baseline interferometers are usually associated with monopulse DF systems used on pulsed (e.g., radar) emitters [56]. Nonmicrowave monopulse DF systems are highly specialized and are generally long-baseline systems.

Single-baseline interferometers can exhibit DOA ambiguities; however, siting may help resolve ambiguities. For example, Figure 5.41 shows a nose-mounted vertical dipole interferometer and a side-mounted cubical quad antenna interferometer sited on the UH-60A helicopter. Both interferometers operate in the VHF band. Airframe shielding resolves fore-aft ambiguity for the dipole interferometer and port-starboard ambiguity for the quad interferometer. Airborne interferometers generally operate above HF due to the limited space for efficient MF and HF siting. Larger aircraft, such as the C-130 and RC-135, can accommodate phase interferometers operating in the high HF band.

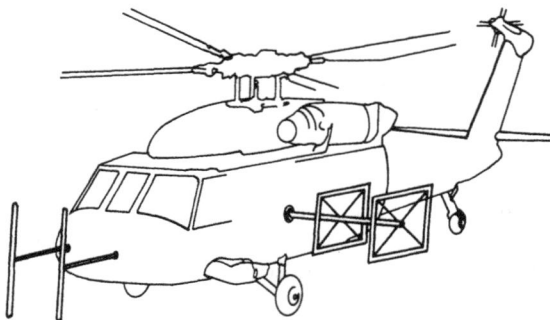

Figure 5.41 Helicopter interferometer configurations.

Long-baseline DF interferometer systems have been used for decades; however, the practical implementation of short-baseline DF interferometers was limited until the advent of efficient, high-speed antenna element switching, multichannel receivers with matched gain and phase and efficient automatic gain control, high-speed analog and digital processing, field-expedient digital data memory, tactical minicomputers, and robust calibration techniques. Advances in broadband RF components also greatly enhanced short-baseline interferometer feasibility.

Doppler and pseudodoppler DF techniques are a variant of direct phase comparison techniques and are included in the Category III systems.

5.4.2 Classes

Interferometer Techniques

Figures 5.42 through 5.45 delineate the phase interferometer classes. Figure 5.42 describes the interferometer class using a two-element antenna–dual-channel receiver. Figures 5.43 and 5.44 pertain to three- and four-element antenna systems, respectively, using dual-channel reception. Figure 5.45 depicts the three-element antenna–triple-channel receiver structure.

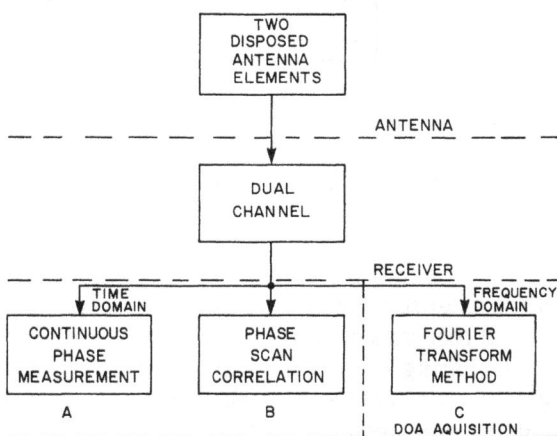

Figure 5.42 Phase comparison DF: two-element antenna–dual-channel receiver class.

The antenna elements are usually disposed in the xy-plane (Figure 2.1) to optimize azimuth AOA measurements. The short-baseline condition constrains all element spacings to no greater than a half wavelength at the highest operating frequency. Two-element arrays are disposed to place the desired azimuthal FOV at, or near, the perpendicular to the baseline between the two elements. Three-element and four-element configurations may be varied to suit the geometry of the host platform, the required FOV, and the capabilities of the DOA computation algorithm, which are often tailored to specific element configurations. For efficient azimuthal FOV, three-element arrays are disposed in either an L-configuration or at the points of an equilateral triangle. Four-element arrays are generally circularly disposed with 90° separation.

The antenna elements are typically electrically short, nonresonant vertical dipoles for airborne and mast-mounted siting, and monopoles for ground-based

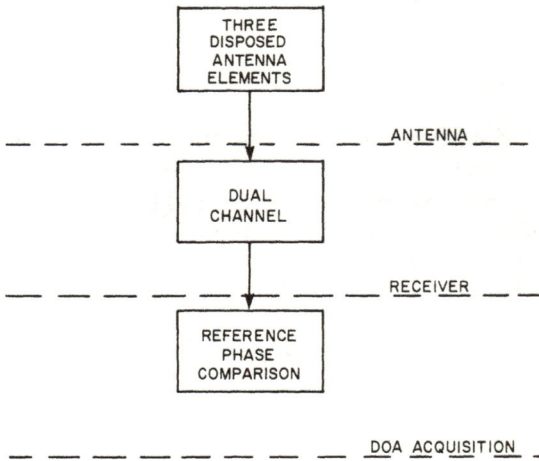

THREE
DISPOSED
ANTENNA
ELEMENTS

— — — — — — — — — — ANTENNA — — —

DUAL
CHANNEL

— — — — — — — — — — RECEIVER — — —

REFERENCE
PHASE
COMPARISON

— — — — — — — — DOA ACQUISITION —

Figure 5.43 Phase comparison DF: three-element antenna–dual-channel receiver class.

sites. Vertical crossed loops that are combined in quadrature are sometimes used as the antenna elements, especially when near-zenith reception and DF are needed.

The multichannel receivers are typically superheterodyne receivers with identical RF-IF channels sharing common synthesized local oscillators. Triple and quadruple frequency conversion are often used. RF calibration via signal injection is used to measure periodically and then remove gain and phase mismatches

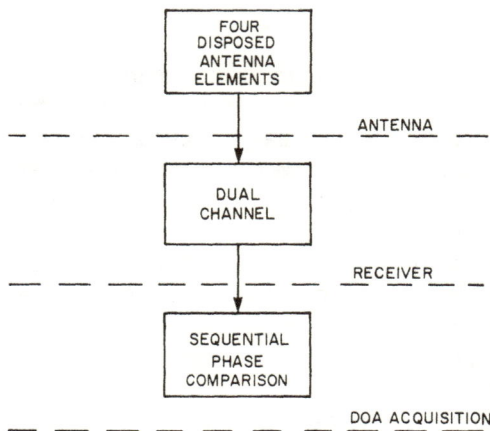

FOUR
DISPOSED
ANTENNA
ELEMENTS

— — — — — — — — ANTENNA — —

DUAL
CHANNEL

— — — — — — — — RECEIVER — —

SEQUENTIAL
PHASE
COMPARISON

— — — — — — — — DOA ACQUISITION

Figure 5.44 Phase comparison DF: four-element antenna–dual-channel receiver class.

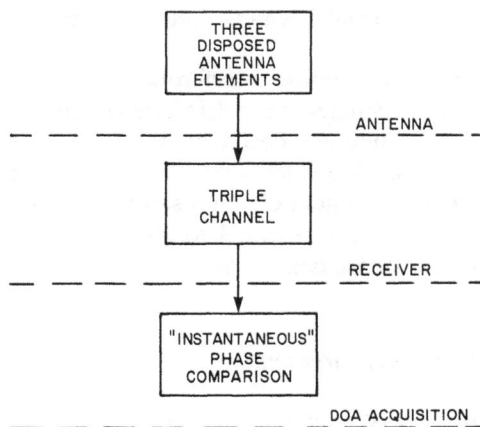

Figure 5.45 Phase comparison DF: three-element antenna–triple-channel receiver class.

between the multiple channels. Periodic calibration alleviates the need for stringent gain and phase tracking between channels.

Pseudodoppler Techniques

Figure 5.46 shows the pseudodoppler DF structure. The pure doppler technique, involving the use of a mechanically rotated antenna, is not practical in the MF through VHF bands.

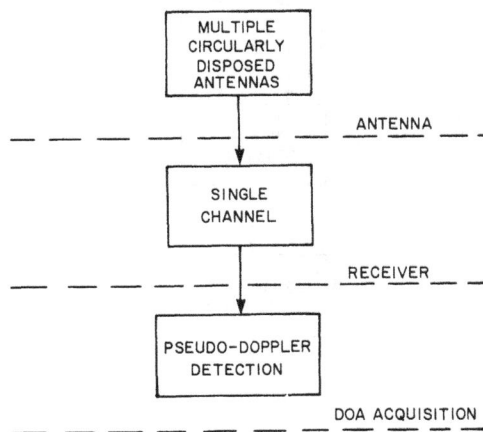

Figure 5.46 Pseudodoppler DF class.

5.4.3 Phase Interferometer: Dual-Element Antenna–Dual-Channel Receiver

Figure 5.42 delineates three basic phase interferometer types using a single baseline and dual-channel receivers. (Single-channel interferometers have been attempted with only limited success. They are based on sequential element sampling and either delay lines or phase-lock loops to "store" the signal phase from one antenna element for comparison with the next element sampled. Practical effectiveness has been limited, and the advent of phase-matched dual-channel receivers has reduced the need for single-channel, phase comparison techniques.)

Type A: Continuous Phase Measurements

A basic phase interferometer is diagrammed in Figure 5.47. The phase differential, δ, between the IF outputs is detected in a phase detector, and the azimuthal bearing angle, φ, is computed knowing δ. Elevation angle is known or assumed.

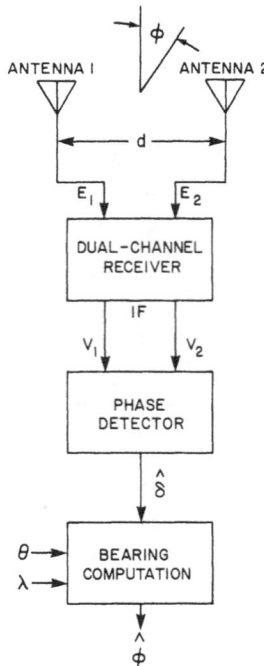

Figure 5.47 Phase comparison using a single-baseline antenna and a dual-channel receiver.

The phase detector may be either analog or digital. Figure 5.48 is a block diagram of an analog approach based on product phase detectors. The integrated outputs of the phase detectors are cosδ (*I*: in-phase) and sinδ (*Q*: quadrature) measures of the phase differential δ. The *I-Q* outputs are either (1) vector summed to provide an analog measure of δ, (2) converted to a digital form in a sine-cosine-to-digital converter, or (3) applied to a CRT to acquire a visual display of δ.

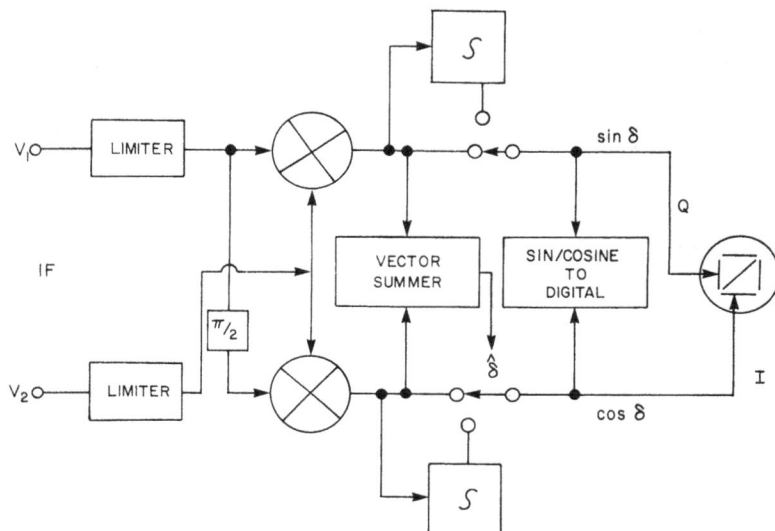

Figure 5.48 Analog phase detector.

Post-detection integration averages across multiple-phase-differential cycles and, hence, provides weak signal enhancement and measurement accuracy improvement. Also, averaging reduces the deleterious effects of signal modulation and multipath interferences.

Phase mismatch between the receiver channels can be a function of frequency across the passband. Hence, for modulated signals, the phase error introduced will be a function of the modulation spectral structure as well as the phase mismatch between channels. For example, an unmodulated carrier located at the center frequency of both channels may experience no phase mismatch between channels. However, a frequency-shift-keyed (FSK) signal with time-variable spectral features that are asymmetrical with respect to the center frequency may experience phase errors unless the channel phase shifts are identical at each of the frequency-shift frequencies. Phase averaging across many frequency-shift transitions reduces phase errors. Averaging also reduces the adverse effects of other asymmetrical modulation formats.

When multipath interference occurs, the resultant phase of the wave produced by the interfering signals varies as the relative phase delays of the signals change. This creates "beat frequencies" that are simply the rates of phase change created by the relative changes in transmission path differences between the interfering waves. If phase averaging is performed across many "beat frequency" cycles, a phase interferometer tends to measure the DOA of the strongest component in the multipath field. This is particularly true if we have only two interfering signals.

Co-channel interference acts like multipath, and it can "capture" a phase interferometer if its strength is greater than the desired signal.

Analog phase measurement accuracy is reduced by signal amplitude and frequency variations. Digital phase measurement techniques are more tolerant of signal variations and can provide a greater degree of accuracy than analog methods.

The strengths and limitations of the single-baseline, dual-channel continuous phase measurement interferometer are as follows:

Strengths

- Nearly instantaneous bearing acquisition
- Relatively simple
- No antenna switching required.

Limitations

- Limited FOV
- Requires knowledge of elevation angle
- Susceptible to interference in the passband.

Type B: Phase Scanning Correlation

Phase scanning correlation is diagrammed in Figure 5.49 [54]. The channel 2 IF output, V_2, is applied to a voltage-variable delay network where the delay $\tau(t)$ is controlled by voltage $v(t)$. The delayed IF of channel 2 is cross-correlated with the undelayed channel 1 IF output V_1. The cross-correlator output is given by

$$F(\tau) = \int_0^T V_1(t)V_2[t - \tau(t)] \, dt \tag{5.88}$$

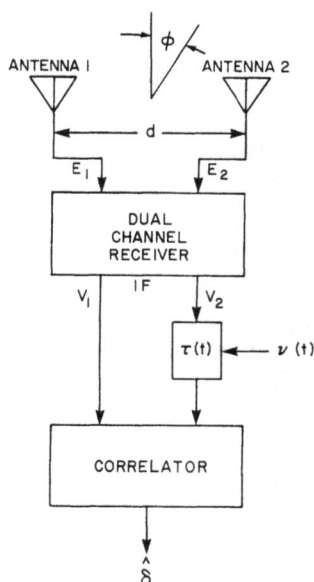

Figure 5.49 Phase scanning correlation DF.

where T is large relative to $\tau(t)$ and is selected based on the precorrelator IF bandwidth W. The value of $\tau(t)$ that maximizes $F(\tau)$ is the maximum likely estimate of δ, the phase differential between the antennas.

Numerous variants exist for the basic approach shown in Figure 5.49. The voltage-variable phase delay $\tau(t)$ could be implemented as an RF function and placed between one antenna and its receiver. Or $\tau(t)$ could be divided into two networks—one a lead phase and the other a lag phase—and placed in opposite channels. Further, $\tau(t)$ could be servoed at either RF or IF to obtain a maximum at the cross-correlator output.

A time-domain cross-correlator improves the SNR; however, as a maximum likelihood estimator (MLE), a cross-correlator is difficult to implement. Also, the maxima of time-domain correlation functions are difficult to measure with high accuracy. In addition, the phase correlator is susceptible to in-band interference.

The strengths and limitations of the phase scanning correlation technique are as follows:

Strengths

- Increased SNR.

Limitations

- Increased bearing acquisition time
- Limited FOV
- Requires knowledge of elevation angle
- Susceptible to interference.

Type C: Fourier Transform Method

Frequency-domain processing is used to measure the phase delay between two spaced antennas [54, 59]. A functional approach is shown in Figure 5.50. Four major steps are involved as follows:

1. Time-domain processing;
2. Frequency-domain processing;
3. Phase-delay computation;
4. Angle-of-arrival computation.

    Time-domain processing involves analog-to-digital (A–D) conversion of the two IF outputs and storage of the digital samples. The rate of conversion is determined by the signal bandwidth prior to the A–D conversion. Window (shaping) functions may be applied to the digital samples after A–D conversion.

    Frequency-domain processing employs Fourier transform algorithms to convert each time-domain sample to complex points or cells in the frequency domain. Fast Fourier transform (FFT) algorithms are used to decrease computation time. Each complex point or cell in the frequency domain has real (r_k) and imaginary (i_k) components where k is the kth frequency cell. The phase in the kth frequency cell is

$$\gamma(k) = \arctan(i_k/r_k) \tag{5.89}$$

and amplitude is

$$V(k) = \sqrt{(r_k)^2 + (i_k)^2} \tag{5.90}$$

The interferometer phase delay for the kth frequency cell is expressed as

$$\gamma(k) = \gamma_1(k) - \gamma_2(k) \tag{5.91}$$

where $\gamma_1(k)$ is the channel 1 phase for the kth frequency cell and $\gamma_2(k)$ is the channel 2 phase for the kth frequency cell.

    Knowing $\gamma(k)$, the angle of arrival $\phi(k)$ for the kth frequency cell may be computed. The overall AOA may be computed by algorithms using the individual

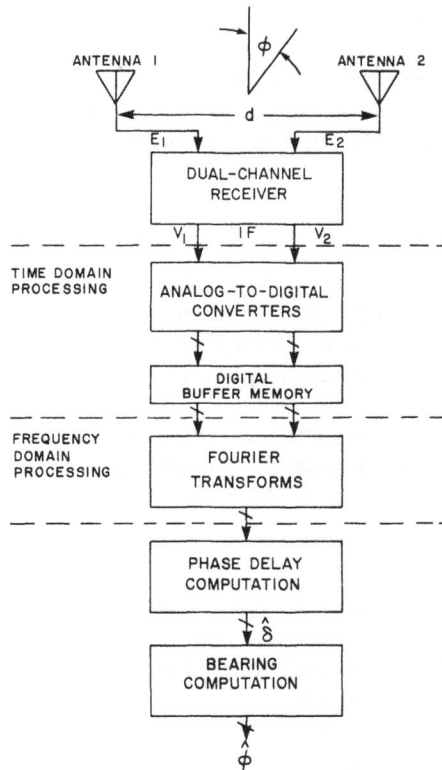

Figure 5.50 Fourier transform phase comparison DF technique.

$\phi(k)$ values. Generally, the angle of arrival for each cell, AOA(k), is computed using $\phi(k)$. Then, if an orthogonal dual-baseline antenna is used, the LOB is computed for each AOA(k) to produce LOB(k). Next, the individual LOB(k) values are processed, sorted, and weighted to obtain an overall LOB estimate. Elevation angle measures are acquired in a similar manner.

The Fourier transform technique has numerous advantages. First, as a digital technique, the Fourier transform reduces the adverse effects of signal amplitude variations. Second, spectral processing can provide 20 to 30 dB of sensitivity improvement. Third, the Fourier transform is compatible with swept-frequency calibration and phase-correction RF signal insertion. Each frequency cell can be corrected for phase errors. Fourth, the Fourier transform is ideally suited for short-duration signal reception and processing because the effective bandwidth is adjustable to match the spectral features of the signal of interest. Fifth, interference and jamming effects are reducible by selective rejection of unwanted spectral components. Sixth, as a digital technique, the Fourier transform output data may be stored and adaptively averaged to improve accuracy. Also, data may be retained and processed for extended time periods without significant degradation. A summary of the dominant strengths and limitations of the Fourier transform technique is as follows:

Strengths

- Robust digital processing in the frequency domain
- Adaptive processing and computational flexibility
- Sensitivity and DOA accuracy improvements
- Short-duration signal operation
- Operation in a dense electromagnetic environment
- Interference and jamming rejection capability.

Limitations

- Relatively complex technique for applications using cooperative transmissions
- Requires knowledge of elevation angle (single baseline)
- Limited FOV (single baseline)
- Possibly limited dynamic range.

The Fourier transform technique is emerging as the dominant system for applications involving short-duration uncooperative transmissions in a dense electromagnetic environment.

5.4.4 Phase Interferometer: Three Antenna Elements and a Dual-Channel Receiver

Phase interferometers with two antenna elements forming one baseline are operationally limited. Two baselines are needed for (1) a full 360° azimuth FOV and (2) elevation angle measurement capability [59]. Figure 5.43 outlines the basic dual-baseline, dual-channel receiver approach; a functional block diagram is shown in Figure 5.51.

Only three antenna elements are needed to form two baselines. The three antenna elements are disposed in the xy azimuth plane in either an L-shape or as an equilateral triangle. (Other configurations may be used, but the computational algorithms may be more complicated.) In Figure 5.51, element 1 is connected to receiver channel 1, whereas channel 2 is alternately switched between elements 2 and 3 by the RF switch. The phase differentials across baselines 1–2 and 1–3 are

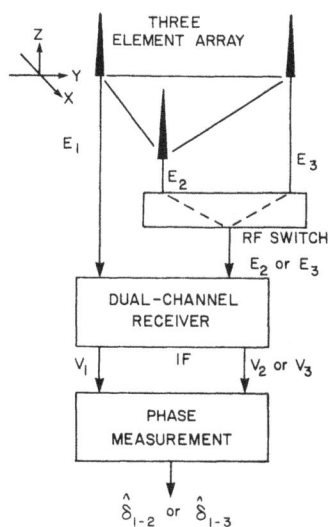

Figure 5.51. Phase comparison using a dual-baseline, dual-channel receiver.

alternately measured as δ_{1-2} and δ_{1-3} where the phase of element 1 is established as the reference phase. Azimuth angle, ϕ, and elevation angle, θ, may be computed from Eqs. (5.76) and (5.77), respectively, where δ_{1-2} corresponds to δ_1 and δ_{1-3} corresponds to δ_2.

Phase measurement methods are analog, digital, or Fourier transform depending on the technical and operational requirements.

The maximum sampling rate between antenna elements 2 and 3 is a primary design consideration, and the selection is often a compromise. The RF switch dwell time on each element (2 and 3) must be long enough to accommodate circuitry and processing response times; however, the sampling rate must be fast enough to obtain measurements on short-duration signals. (A typical short-duration signal is a frequency-hopper that dwells in the phase interferometer passband for 10 milliseconds.)

Pertinent strengths and limitations of the three-element, dual-channel interferometer are as follows:

Strengths

- Measurement of both azimuth and elevation angles
- Improved FOV relative to a single baseline
- Reduced antenna scattering and coupling errors (one antenna of the three is switched off at all times).

Limitations

- Requires antenna sampling
- Increased DOA acquisition time (caused by sampling).

5.4.5 Phase Interferometer: Four Antenna Elements and a Dual-Channel Receiver

Using a dual-channel receiver, sequential phase comparisons may be obtained across the six baselines formed by four disposed antenna elements. Figure 5.44 depicts the system structure, and Figure 5.52 shows a functional block diagram [59]. The four antenna elements are arrayed with each element at the corner of a square with diagonal separation d. The common practice is to use the phase differential across the diagonal baselines 1–3 and 2–4. In Figure 5.52, with the phase reference at the intersection of the diagonal baselines, the antenna element outputs may be written as

$$E_1 = \cos(\omega_c t + \beta \cos\phi) \tag{5.92}$$

$$E_2 = \cos(\omega_c t + \beta \sin\phi) \tag{5.93}$$

$$E_3 = \cos(\omega_c t - \beta \cos\phi) \tag{5.94}$$

$$E_4 = \cos(\omega_c t - \beta \sin\phi) \tag{5.95}$$

where $\beta = (\pi d/\lambda) \sin\theta$ and ϕ is the azimuth angle of arrival and θ is the elevation angle of arrival.

Sequential phase measurements are made across diagonals 1–3 and 2–4, and receiver channel interchange is performed on each baseline. The phase samples are expressed as

$$\delta_{1-3} = \chi + 2\beta \cos\phi \tag{5.96}$$

$$\delta_{3-1} = \chi - 2\beta \cos\phi \tag{5.97}$$

$$\delta_{2-4} = \chi + 2\beta \sin\phi \tag{5.98}$$

$$\delta_{4-2} = \chi - 2\beta \sin\phi \tag{5.99}$$

where χ is the phase mismatch between the two receiver channels and is assumed to remain constant during the four sequential phase measurements. The phase mismatch χ is removed by subtraction of the phase samples, which provides

$$(\delta_{1-3} - \delta_{3-1}) = 4\beta \cos\phi \tag{5.100}$$

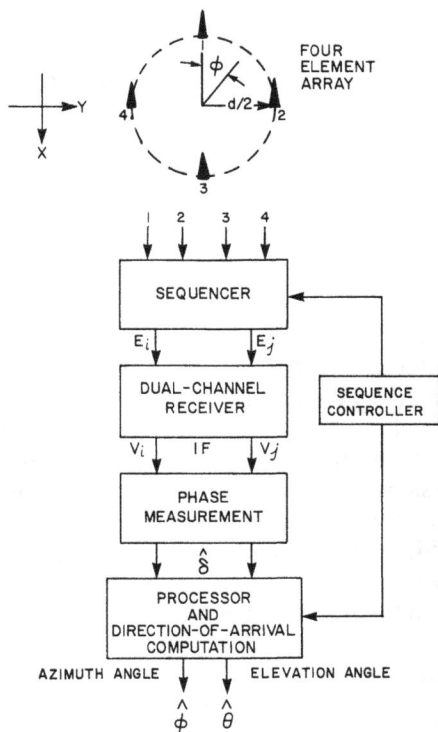

Figure 5.52 Sequential phase comparison DF technique.

$$(\delta_{2-4} - \delta_{4-2}) = 4\beta \sin\phi \qquad (5.101)$$

Processing of the phase-delay estimates obtained from Eqs. (5.100) and (5.101) yields azimuth and elevation angle information. The four-element/dual-channel phase interferometer has been widely used for MF and HF propagation research [60–64].

The dominant strengths and limitations of the four-element/dual-channel phase interferometer technique are as follows:

Strengths

- Full 360° azimuthal FOV
- Matched channels not required due to use of channel interchange
- Six baselines available for frequency coverage requirements.

Limitations

- DOA information acquisition time is increased and sensitivity is reduced by the sequential sampling and channel interchange
- Sampling rate is a function of the phase modulation on the incident signal.

5.4.6 Phase Interferometer: Three Antenna Elements and a Triple-Channel Receiver

Instantaneous DF over a full azimuthal FOV requires that a phase interferometer have three disposed antennas with each antenna continuously connected to a receiver channel. Figure 5.45 delineates the approach; a functional block diagram is shown in Figure 5.53. No antenna switching is required; hence, all phase measurements and DOA computations are made in "real time," subject to the time constraints imposed by signal processing and parameter computation.

The three antennas are typically deployed to form an equilateral triangle as illustrated by Figure 5.54, which shows the geometry of the antenna configuration and the signal direction of arrival. Letting $\zeta = 60°$ in Figure 5.54, the differential phases are given by the following expressions:

$$\delta_{1-2} = (2\pi d/\lambda) \cos\phi \sin\theta \qquad (5.102)$$

$$\delta_{3-1} = (2\pi d/\lambda) \cos(\phi - 120°) \sin\theta \qquad (5.103)$$

$$\delta_{2-3} = (2\pi d/\lambda) \cos(\phi - 240°) \sin\theta \qquad (5.104)$$

The phase differentials are used to find the elevation angle and the three azimuth angles of arrival referenced to the three baselines. The equations are as follows:

$$\theta = \arcsin\{(\sqrt{\delta_{1-2}^2 + \delta_{3-1}^2 + \delta_{2-3}^2})/[(2\pi d/\lambda)(\sqrt{3/2})]\} \qquad (5.105)$$

$$\phi_{1-2} = \arctan\{(2\delta_{2-3} + \delta_{1-2})/[(\sqrt{3})(\delta_{1-2})]\} \qquad (5.106)$$

$$\phi_{1-3} = \arctan\{(-2\delta_{3-1} - \delta_{1-2})/[(\sqrt{3})(\delta_{1-2})]\} \qquad (5.107)$$

$$\phi_{2-3} = \arctan\{(\delta_{2-3} - \delta_{3-1})/[(-\sqrt{3})(\delta_{2-3} + \delta_{3-1})]\} \qquad (5.108)$$

The three values for ϕ_{i-j} obtained from Eqs. (5.106) through (5.108) may be vector summed to obtain an expression for the azimuth DOA, ϕ, referenced to the coordinate system shown in Figure 5.54. The expression is

$$\phi = \arctan\{[\cos(\phi_{1-2}) + \cos(\phi_{1-3}) + \cos(\phi_{2-3})]/[\sin(\phi_{1-2}) \qquad (5.109)$$
$$+ \sin(\phi_{1-3}) + \sin(\phi_{2-3})]\}$$

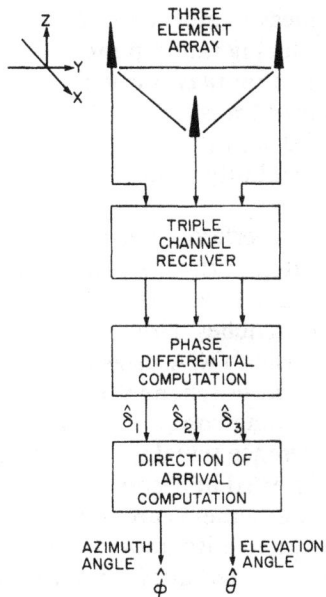

Figure 5.53 Instantaneous phase comparison DF technique.

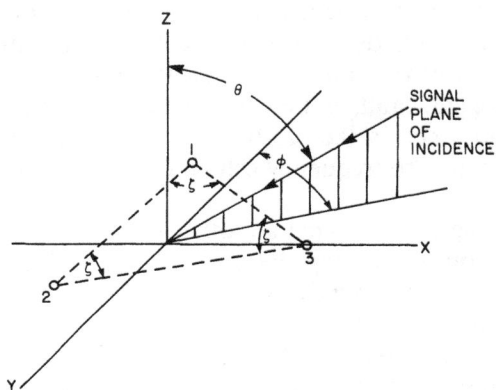

Figure 5.54 Geometry of triple-channel receiver configuration.

Only two of the three phase measurements are required; however, inclusion of all three increases accuracy by reducing random measurement errors.

Phase-differential computation may use analog, digital, or digital with Fourier transform methods. The Fourier transform method is favored due to advantages such as coherent averaging of data, interference and jamming mitigation, calibration and compensation compatibility, enhanced DOA accuracy, and computational flexibility.

The effectiveness of phase interferometer systems is a function of the real-time knowledge of the phase mismatch in the system. Phase matching and tracking are inherently difficult in triple-channel receivers. The sharing of common local oscillators negates major phase mismatches. However, additional real-time phase mismatch data are needed for precision DOA measures. Channel interchange can provide phase mismatch data; however, channel interchange compromises response time and adds a lossy RF switching component prior to the receiver. A common approach is to insert an RF reference signal periodically at the operating frequency. The RF calibration signal is inserted at a point symmetrical with the antenna element phase centers; the antenna elements are isolated from the receiver during the calibration signal insertion. (Antenna isolation reduces the effects of antenna interaction, signal cross-coupling, and radiation that could compromise DF system covertness.) The amplitude of the RF calibration may be either a preset nominal value or at a level approximately equal to the signal of interest. The amplitude and phase differentials between channels are measured, stored, and used in subsequent phase measurements to remove mathematically receiver channel amplitude and phase mismatch errors. The RF signal insertion and measurement requires a fraction of a second and is repeated every several minutes. Therefore, the DF system response time is not seriously impaired.

A narrowband RF calibration signal centered in the passband cannot define phase errors on the sidebands. A stepped or swept RF calibration signal is necessary for full passband phase-error definition. A swept RF calibration signal is ideally suited for use with Fourier transform phase measurements. As the RF calibration signal is swept over the passband, an FFT (or DFT) is performed to define the phase error for each frequency cell in the passband. This cellular phase error is used to adjust the phase error in each frequency cell and, hence, remove the effects of system-induced phase errors.

Antenna scattering and coupling are other sources of error because all three antennas are actively on-line during the phase-differential measurement process. These errors are systematic and relatively constant with time and site effects. Therefore, site calibration data contain information on antenna and scattering errors, and, if calibration is performed at a "Class A" site that is relatively free from reradiators and terrain features, the site calibration data tend to delineate antenna scattering and coupling errors. Correction tables can be generated and, possibly, stored in the DOA computation circuitry for error compensation. The pattern correlation technique discussed in Section 6.3.1 may also be used.

The dominant strengths and limitations of the three-element–triple-channel phase interferometer are as follows:

Strengths

- Instantaneous DOA information
- Full azimuthal FOV
- Azimuth and elevation angles are computed
- No RF switching required; leads to increased sensitivity and reduced response time
- Compatible with Fourier transform processing
- Relatively tolerant of modulation
- Three baselines are available for phase-delay averaging.

Limitations

- Requires robust, high-speed computational capability
- Needs injection signal calibration and site calibration data for error reduction
- Relatively complex circuitry.

5.4.7 Pseudodoppler (Quasidoppler)

In a pure doppler DF system, the received frequency from a moving antenna experiences a frequency shift created by the doppler effect. Figure 5.55 shows a vertical monopole rotating in a circle with radius r. Let the received signal from an imaginary stationary antenna at the center point c be

$$E = E_0 \sin(\omega_0 t) \tag{5.110}$$

where $\omega_0 = 2\pi f_0$ and f_0 is the received frequency. The received signal at point A is

$$E_A = E_0 \sin[\omega_0 t + (2\pi r/\lambda)(\cos\phi)] \tag{5.111}$$

If an antenna is rotated at a uniform rate ω_r in a circle,

$$\phi = \omega_r t \tag{5.112}$$

and

$$E_A = E_0 \sin[\omega_0 t + (2\pi r/\lambda)(\cos\omega_r t)] \tag{5.113}$$

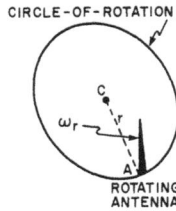

Figure 5.55 Rotating antenna parameters.

When E_A is applied to a sinusoidal phase detector, the output is

$$V_P = \sin[(2\pi r/\lambda)(\cos\omega_r t)] \tag{5.114}$$

$$= \sin[(R)(\cos\omega_r t)] \tag{5.115}$$

where $R = 2\pi r/\lambda$ is half the aperture of the circle in radians. Using a shifted time base, V_P is a Bessel function series of odd harmonics given by

$$V_P = 2[J_1(R) \sin\omega_r t + J_3(R) \sin3\omega_r t$$
$$+ J_5(R) \sin5\omega_r t \ldots J_{2n+1}(R) \sin(2n+1)\omega_r t + \ldots] \tag{5.116}$$

Therefore, the output of a continuously rotating antenna is a series of audio tones with frequencies at odd harmonics of the rotation rate and with amplitude expressed by Bessel functions of the first kind [65]. The signal experiences a peak doppler shift of $\omega_r r/\lambda$. The signal bearing is tangent to the circle of rotation at the maximum doppler point and is in the direction of the antenna movement. In a complete revolution, the instantaneous frequency shift varies from $(\omega_0 + \omega_r r/\lambda)$ to $(\omega_0 - \omega_r r/\lambda)$ where f_0 is the operating frequency. The tangential velocity, v_t, of the rotating antenna is derived as follows. Because

$$\omega_r r/\lambda = f v_t/c = f_s \tag{5.117}$$

then

$$v_t = f_s c/f \tag{5.118}$$

where f_s is the doppler frequency in Hz.

Equation (5.118) may be used to show that the mechanical rotation of an antenna to obtain DF doppler is impractical at frequencies below the UHF band.

For example, assume that $f_s = 100$ Hz is the desired maximum doppler frequency, and the operating frequency is at 3.0 MHz. Then

$$v_t = 100 \exp8/\exp6 = 10{,}000 \text{ m/s}$$

which is an impractical mechanical rotation velocity. Therefore, the mechanical rotation of an antenna is simulated by placing a number of fixed omnidirectional antennas in a circular array and scanning the antennas in sequence by RF switching. This cyclical differential measurement of phase constitutes a pseudodoppler DF approach. Basic early work [66] on this technique described it as a "commutated-aerial direction-finding system" in which the phase of the signal is sequentially sampled at a number of points, by a single receiver, and the signal direction of arrival is computed from the measured relative phase. Figure 5.56 is a functional diagram of the pseudodoppler DF technique. The pseudodoppler imposed on the signal at RF is detected at IF by a frequency discriminator and processed to obtain ϕ.

The antenna arrays usually have from 12 to 30 dipoles or monopoles spaced around the circumference of a circle with a diameter of $\lambda/2$ or less.

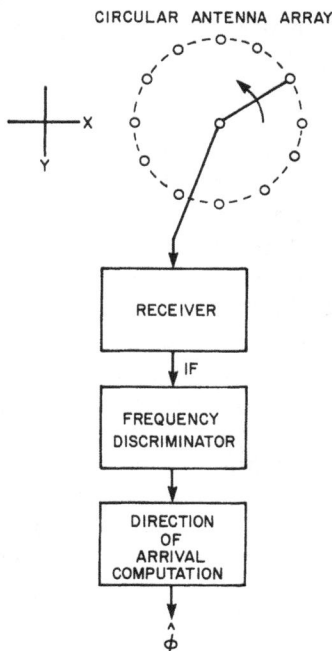

Figure 5.56 Block diagram of pseudodoppler DF system.

The pseudodoppler principle of operation is simple, but it requires numerous design trade-offs related primarily to reducing the interference created by antenna square-wave sampling [67]. Periodically sampled antennas provide an output with numerous AM sidebands around the received signal. Also, the FM created by the periodic frequency shift of the pseudodoppler creates sidebands. The AM and FM sidebands are produced by interdependent processes; therefore, they interact. Each AM sideband created by antenna sampling is frequency modulated by the array rotation. In a dense signal environment, the AM and FM effects can generate multiple sidebands from undesired co-channel or adjacent channel signals. The DF receiver cannot discriminate among multiple signals if their sidebands overlap. The FM information exists on an unwanted sideband, and the DF receiver may phase-lock on the sideband of an unwanted signal. Then, the DF processor may present stable but inaccurate DOA information.

When a pseudodoppler DF system functions as a combined intercept and DF system, the antenna switching noise can degrade the desired signal intelligence. A separate intercept receiver may be necessary.

The antenna switching interference leads to the consideration of design trade-offs as discussed below [67]:

1. *Presampling filtering:* RF filtering prior to the sampler reduces the number of signal sidebands and, hence, decreases the probability of undesired switching sidebands. However, this practice decreases the bearing acquisition time by reducing the effective signal power available for processing. Also, reduced signal bandwidth can increase the tuning time and frequency set-on. Presampling filtering may adversely effect both probability of detection and DOA acquisition time; therefore, presampling filtering is not an attractive trade-off.

2. *Sampling waveform smoothing:* Smoothing of the square-wave sampling waveform by filtering reduces the number and magnitude of the undesired AM sidebands. The trade-off is that the smoothing also "softens" the desired FM phase transitions and reduces the accuracy of frequency-shift detection.

3. *Decreased sampling rate:* A decreased antenna sampling rate reduces AM sideband bandwidth and interference; however, lowering the sampling rate lowers the FM modulation index, which, in turn, reduces DOA measurement accuracy. A lowered sampling rate increases DOA acquisition time and degrades performance on short-duration signals.

4. *Reduce the number of antenna elements:* A reduction in antenna elements lowers the number of switching transients per unit time. An adverse effect is degradation of frequency-shift resolution and, hence, reduced DOA accuracy.

For uncooperative emitters, the design trade-offs are generally nonoptimal; therefore, the best application for the pseudodoppler DF technique is on cooperative emitters in a well-controlled electromagnetic environment. Further, the relatively large antenna system generally requires a fixed, surface-based site [68]. (However, mobile pseudodoppler DF systems have been implemented and are successfully employed [69].) The pseudodoppler DF technique is extensively used in the marine (150- to 174-MHz) and avionics (115- to 144- and 225- to 400-MHz) bands where the emitters are generally "cooperative," the electromagnetic environment is not crowded, and channel spacing is well controlled.

The pseudodoppler DF technique is a function of elevation angle of arrival. The instantaneous pseudodoppler frequency shift is proportional to $\sin\theta$ where θ is the elevation angle. Accuracy decreases as elevation angle decreases because the instantaneous frequency shift decreases. The frequency shift *versus* elevation angle is deterministic; therefore, a coarse measure of elevation angle may be made by measuring frequency shift and using the frequency shift *versus* elevation angle function to assess elevation angle of arrival.

The strengths and limitations of the pseudodoppler DF technique are as follows:

Strengths

- Relatively simple operating principle
- Straightforward implementation
- Simple computational algorithms
- Relatively rapid bearing acquisition
- Coarse measure of elevation AOA possible under well-controlled conditions with calibrated performance.

Limitations

- Antenna sampling introduces signal distortion and erroneous bearings
- Operation in a dense electromagnetic environment compromises performance
- Accuracy decreases as elevation angle decreases; best performance on near-horizon transmissions
- Requires relatively large antenna system restricting siting flexibility and host platform compatibility.

5.4.8 Instrumental and Observational Errors

Phase Interferometers

Phase interferometers are subject to four major instrumental error sources as discussed below:

1. Antenna scattering and coupling;
2. Thermal noise;

3. Multichannel phase mismatch and mistracking;
4. Computational approximations.

As discussed in Section 5.3.6, errors introduced by antenna scattering and coupling are deterministic; therefore, these errors can be defined by site calibration and negated by the DF algorithm.

The primary effect of thermal noise is to introduce errors into the phase measurement process. In the presence of thermal noise, any phase-differential measurement becomes an estimate and may be analyzed using parameter estimation theory [54–56, 70].

Measured angles of arrival ϕ and θ are only estimates $\hat{\phi}$ and $\hat{\theta}$ in the presence of error sources. The rms or standard deviation, s, is given by

$$s = \sqrt{E[(\hat{\phi} - \phi)^2]} \tag{5.119}$$

where the expected value

$$E[(\hat{\phi} - \phi)^2] = \sigma_\phi + b_\phi^2 \tag{5.120}$$

In Eq. (5.120), σ_ϕ is the variance of ϕ and b_ϕ is the error bias given by

$$b_\phi = E[\hat{\phi}] - \phi \tag{5.121}$$

The estimated rms bearing error, $\hat{\phi}_\epsilon$, may be written as

$$\hat{\phi}_\epsilon = \frac{(\hat{\delta}_\epsilon)}{(2\pi d/\lambda)(\cos\phi \, \sin\theta)} \tag{5.122}$$

where
$\delta\epsilon$ = the estimated rms phase error;
d = the baseline spacing;
λ = the wavelength;
ϕ = the azimuth angle of arrival;
θ = the elevation angle of arrival.

The rms phase error is a function of the SNR. Figure 5.57 depicts a phasor diagram with a signal, E, at a phase angle δ, where δ is perturbed by noise, n_t, to produce a phase error, $\hat{\delta}_\epsilon$. The noise n_t, which is assumed to be random, has components that are in phase (n_i) and in quadrature (n_q) to the signal vector, E. The quadrature component, n_q, contributes to the phase error, δ_ϵ. Because the total noise power is

$$n_t^2 = n_i^2 + n_q^2 \qquad (5.123)$$

and $n_i = n_q$ by assumed symmetry,

$$n_t^2 = 2n_q^2 \qquad (5.124)$$

Therefore,

$$n_q = \sqrt{\frac{n_t^2}{2}} \qquad (5.125)$$

For a relatively high SNR,

$$\tan\delta_\epsilon \approx \delta_\epsilon = n_q/E \qquad (5.126)$$

With $E = \sqrt{S}$, where S is the total signal power,

$$\hat{\delta}_\epsilon = \left(\sqrt{\frac{n_t^2}{2}} \right)\left(\frac{1}{\sqrt{S}} \right)$$
$$= \frac{1}{\sqrt{2SNR}}, \text{ radians} \qquad (5.127)$$

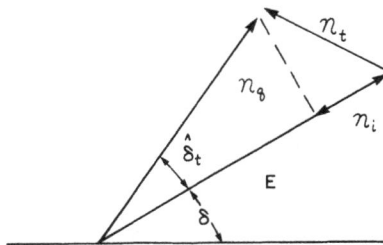

Figure 5.57 Signal phasor with noise corruption.

Because n_t^2 equals the total power, N, $\hat{\delta}_\epsilon$ is the phase error for one phase measurement. In a phase interferometer, two independent phase measurements are made to obtain the final differential phase. In this case,

$$\hat{\delta}_{\epsilon t} = 1/\sqrt{\text{SNR}} \tag{5.128}$$

assuming decorrelated noise signals and equal SNR values. With the reference phase at the center point of the baseline, the estimated azimuth angle of arrival, $\hat{\phi}_\epsilon$, in Eq. (5.122) can be expressed as

$$\hat{\phi}_\epsilon = \frac{1}{(\pi d/\lambda)(\cos\phi)(\sin\theta)\sqrt{\text{SNR}}} \quad , \text{radians} \tag{5.129}$$

Figure 5.58 is a plot of estimated azimuth angle of arrival error *versus* SNR for a d/λ value of 0.5 with $\phi = 0°$ and $\theta = 90°$.

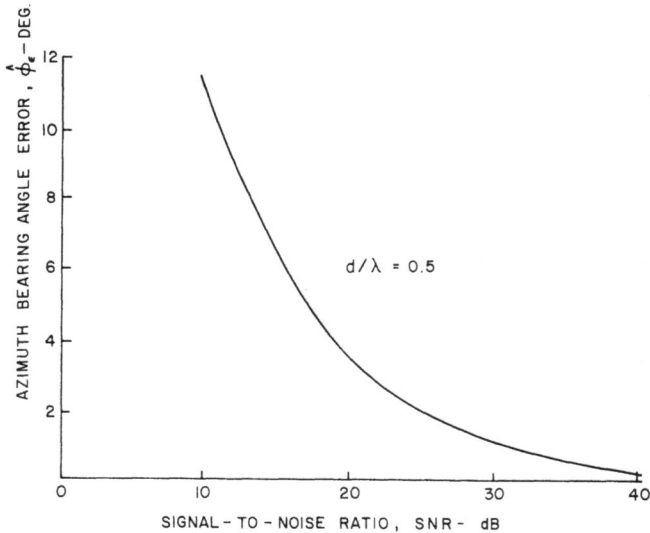

Figure 5.58 Phase comparison bearing error *versus* SNR.

The form of the bearing error expression may vary as a function of the specific phase measurement technique used. For example, phase measurements using cross-correlation techniques [54] produce a different form of $\hat{\phi}_\epsilon$ that includes the effects of correlation time on SNR. The digitized measures of phase used for cross-correlation may be time averaged to reduce thermal noise effects.

Multichannel phase mismatch and mistracking errors are deterministic; however, these errors are a function of time, temperature, signal conditions, and operational effects such as maintenance. Periodic test signal insertion is required at a rate consistent with the rate of change of phase mismatch and with minimal impact on DOA acquisition time. Clearly, a design trade-off is required. A critical receiver function is the automatic gain control (AGC). The AGC for multichannel receivers should be derived from a common control circuit for best phase and amplitude tracking.

Algorithms are used in DOA computations; in most cases, these algorithms approximate certain functions. For example, a truncated form of the Taylor series expansion of the arcsine function is commonly used in DOA algorithms to reduce computation time. Residual DOA errors result from the algorithm truncation; however, these errors are predictable and may be reduced by linear interpolation techniques. Generally, algorithm errors are relatively small if correction procedures are used and may be ignored if the correction schemes are robust.

Pseudodoppler

Major instrumental error sources in a pseudodoppler DF system are thermal noise, circuitry frequency stability, and circuitry phase linearity and stability. (Antenna sampling also creates errors; however, these errors are inherent in the pseudo-doppler DF technique and are often irreducible.)

Thermal noise effects are comparable to those for a conventional FM or PM receiving system. The SNR should be high to ensure good amplitude limiting action prior to the phase discriminator. If the signal of interest fades or ceases to transmit, external noise or interference may "capture" the receiver and introduce direction-of-arrival errors.

Frequency stability is required to ensure consistent operation at the center frequency of the phase discriminator. A frequency offset will result in a phase measurement error. Figure 5.59 shows a compensation technique for frequency off-set, Δf.

Phase linearity and stability directly affect direction-of-arrival measurement accuracy. The prediscriminator amplifier-limiter is a critical stage in the receiver. Typical amplifier-limiter stages introduce considerable phase shift as the input signal amplitude varies. Constant-transmission, phase-limiting amplifiers are desired because this type of amplifier-limiter introduces minimal phase error as a function of input signal amplitude. At VHF, a typical constant-transmission, phase-limiting amplifier introduces less than $\pm 0.5°$ phase deviation over a 40-dB input signal amplitude range.

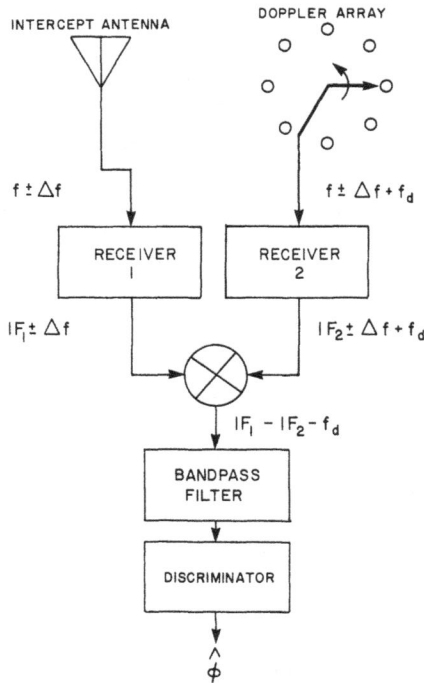

Figure 5.59. Block diagram of pseudodoppler dual-receiver frequency compensation technique.

5.5 CATEGORY IV: TIME DIFFERENCE OF ARRIVAL

Historically, time-difference-of-arrival techniques have not been used with small-aperture DF systems. TDOA techniques generally use long baselines of multiple wavelength; however, improvements in time-interval measurement techniques have enhanced the potential for effective short-baseline TDOA DF. Direction-of-arrival information is obtained from the time difference between the arrival time of a signal at two disposed antennas on a baseline of distance d. (The signal is assumed to be pulsed.)

In Figure 2.8, the TDOA, t_d, is calculated from

$$t_d = (d/c) \sin\phi \sin\theta \tag{5.130}$$

where c is the velocity of light and ϕ and θ are the azimuth and elevation angles, respectively. For spacing d in meters and time in nanoseconds,

$$t_d = 3.33d \sin\phi \sin\theta \tag{5.131}$$

Time difference t_d is independent of the operating frequency; therefore, the $d/\lambda \leq$ 0.5 criteria is redefined for TDOA systems. Specifically, the state of the art in time-interval measurements determines how short a TDOA baseline can be. In the 1960s, short-baseline TDOA systems were measured in kilometers. In the 1970s, short-baseline TDOA systems operated with 100-m baselines. In the 1980s, the baselines were in the ten's of meters and became operationally compatible with mobile platforms (aircraft, ships, surface vehicles, *et cetera*) and mast-mounting. Limited-area, tactical deployment began. In the 1990s, the use of short-baseline TDOA DF systems is expected to grow as a direct function of the increase in time-interval measurement accuracy and resolution.

The azimuth angle of arrival is obtained from Eq. (5.131) and is expressed as

$$\phi = \arcsin[(t_d)/(3.33d\,\sin\theta)], \text{ radians} \tag{5.132}$$

A three-element antenna and triple-channel receiver are used to provide separate, independent measures of azimuth angle, ϕ, and elevation angle, θ, without DOA ambiguity. Figure 5.60 shows the geometry for a three-element, dual-baseline TDOA system. Azimuth angle is computed from

$$\phi = \arctan[(t_{d1})/(t_{d2})] \tag{5.133}$$

and the elevation angle is obtained from

$$\theta = \arcsin\{(c/d)(\sqrt{t_{d1}^2 + t_{d2}^2})\} \tag{5.134}$$

where c is the velocity of light.

The TDOA technique requires that one antenna be a reference antenna that "starts" the time-interval measurement. Thus, if the incident signal arrives at the nonreference antennas first, no time-interval measurement can be initiated. This problem is resolved by inserting known time delays in the nonreference channels and selecting the time delays to be d/c seconds or slightly longer.

Implicit in Equation (5.132) is the assumption that the distance to the RF source is much greater than the baseline distance, d. For short-baseline systems, this is a good assumption. However, for some long-baseline TDOA systems, this assumption may not be valid, and DOA algorithms other than Eqs. (5.133) and (5.134) must be used.

Major design issues for a short-baseline TDOA system are the trade-offs between baseline distance, time-interval measurement (TIM) resolution and error, and desired DOA accuracy. Generally the first design step is to select a maximum baseline distance compatible with the host platform and deployment situation. Next, DOA error is computed based on (1) the best resolution of the TIM without

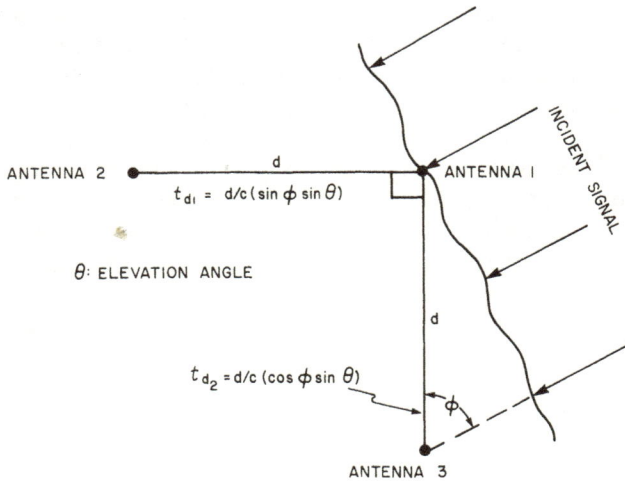

Figure 5.60 Time difference-of-arrival DF configuration.

error sources and (2) TIM performance in the presence of error sources. The error in azimuth angle, ϕ_ϵ, as a function of TIM error is expressed by

$$\phi_\epsilon = [(17.2t_{de})/(d\cos\phi)] \tag{5.135}$$

where t_{de} is the time measurement rms error in nanoseconds, d is in meters, and ϕ_ϵ is in degrees. The elevation angle is assumed to be 90°. A normalized plot of ϕ_ϵ/t_{de} *versus* azimuth angle ϕ is given in Figure 5.61 for three representative baseline spacings.

The current state of the art in time-interval measurement accuracy is less than ± 1 ns. Using a conservative value of ± 0.5 ns accuracy, the minimum ϕ_ϵ values for 1- and 10-m baselines are 8.6° and 0.86°, respectively. Clearly, the use of a 1-m baseline can produce unacceptable DOA errors, but the errors for a 10-m baseline are reasonable for many applications if the azimuth angle does not exceed about 60°.

Imperfect conditions further degrade TIM accuracy from its basic resolution capability. Major degradation factors are thermal noise, which reduces the SNR, and signal dispersion, which reduces the pulsed signal leading and trailing edge rise times. Reference [56] presents a detailed analysis of TDOA system performance in the presence of a finite SNR and incident signal rise and fall time. (Some time-interval measurements are made on the trailing edge of the pulse.) TIM errors are also discussed in References [54], [55], and [70].

Signal pulse dispersion can be caused by both external sources, such as signal multipath, and internal sources, such as nonlinear video detection and nonoptimal receiver bandwidth. The rms error for a time-interval measurement on the leading edge of a pulse is given by

$$t_{de1} = (t_r)/\sqrt{2\text{SNR}} \tag{5.136}$$

where t_r is the video pulse rise time and SNR is the video signal-to-noise ratio. Linear video detection and a large SNR are assumed. At frequencies below microwave, t_r will probably be degraded by external sources such as multipath, transmission path pulse dispersion, and local reradiation and scattering of the incident RF signal. However, in some situations, t_r may be degraded by the DF receiver video bandwidth, B_v. The RF and IF bandwidths are generally adequate for faithful pulse transmission. Signal rise time as a function of video bandwidth is given by

$$t_r = 0.35/B_v \tag{5.137}$$

Two independent time-interval measurements are made to arrive at the TDOA across a baseline. The two combined measurements are improved by the factor $\sqrt{2}$ to obtain the final, total rms error, t_{de}, as

$$t_{de} = t_r/\sqrt{\text{SNR}} \tag{5.138}$$

or

$$t_{de} = (0.35)/[(B_v)(\sqrt{\text{SNR}})] \tag{5.139}$$

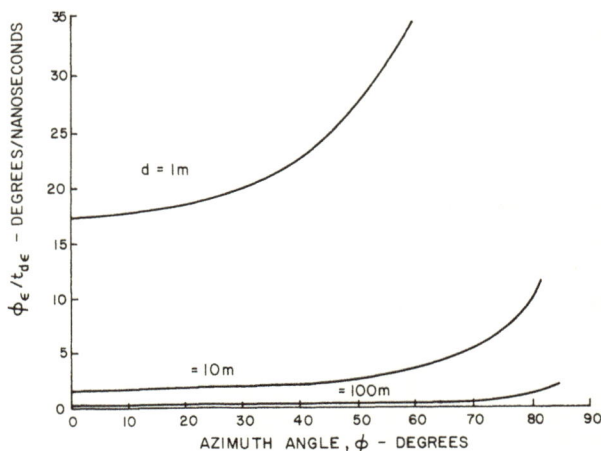

Figure 5.61 Normalized bearing error *versus* azimuth angle for a TDOA DF system.

For typical values of B_v and SNR encountered in practice, t_{de} exceeds the basic resolution accuracy of present TIM methods. If the pulsed RF source is repetitive, e.g., a radar, multiple time-interval measurements can be performed and averaged. If N time-interval measures are taken and averaged, t_{de} is reduced by a factor $\sqrt{N}$. Unfortunately, at MF through VHF, natural and man-made pulsed RF sources are not generally orderly and repetitive if cooperative emitters are excluded from consideration. Overall, the practical requirement for numerous averaged time-interval measurements to reduce error to a satisfactory level compromises the utility of TDOA DF systems in the MF through VHF bands, where repetitive RF pulsed sources are rare. One notable exception is the relatively large number of sferic pulses associated with a lightning flash. The repetitive nature of sferics enhances the feasibility of short-baseline sferic TDOA DF systems [71, 72].

A functional block diagram of a dual-baseline, three-channel receiver TDOA DF system is shown in Figure 5.62. Antenna 1 provides the reference signal that starts the time-interval measurements, whereas antennas 2 and 3 stop the time-interval measurements. The time delays in channels 2 and 3 are slightly larger than d/c seconds to ensure "positive" time-interval measures and resolve DOA ambiguity.

Conceptually, short-baseline TDOA systems are simple and straightforward. Practically, TDOA systems are difficult to implement. Instrumental error sources, such as channel balance and time-interval measurement stability, must be established and maintained at very high performance levels. Channel effects include antenna scattering and coupling, which can introduce both pulse time dispersion and amplitude unbalance. Amplitude unbalance between channels results in time-interval measurement errors. Therefore, adaptive measurement thresholding [73] in the time-interval measurement circuitry is essential for high accuracy.

The strengths and limitations of the TDOA short-baseline DF technique are as follows:

Strengths

- Conceptually simple
- Independent of frequency (in theory)
- Rapid DOA acquisition
- Low power requirements
- Low operator proficiency needed.

Limitations

- Modulation dependent; operates on pulsed sources
- Channel matching and tracking required

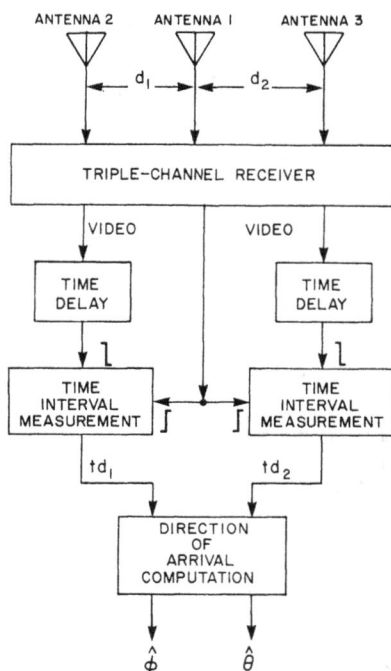

Figure 5.62 Block diagram of a TDOA DF system.

- Requires state-of-the-art time-interval measurements
- Broadband devices and components needed
- Wide bandwidth requirement increases susceptibility to interference and jamming
- Performance degrades rapidly as SNR decreases
- Performance is compromised by external and internal pulse dispersion
- Relatively clean siting and host platform are needed to minimize pulse dispersion
- Periodic calibration is necessary to maintain accuracy.

REFERENCES

1. Bond, D.S., *Radio Direction Finders,* New York: McGraw-Hill Book Co., 1944.
2. Keen, R., *Wireless Direction Finding,* 4th Ed., London: Iliffe and Sons, Ltd., 1947.
3. Gething, P.J.D., "High Frequency Radio Direction Finding," *Proc. IEE,* Vol. 113, No. 1, January 1966, pp. 49–61.

4. Terman, F.E., *Electronic and Radio Engineering,* New York: McGraw-Hill Book Co., 1955, Chapter 22, pp. 803–825.
5. Boyd, J.A., D.B. Harris, D.D. King, and H.W. Weitch, Jr., eds., *Electronic Countermeasures,* Los Gatos, CA: Peninsula Publishing Co., 1978, Chapter 10.
6. *Radio Direction Finding,* U.S. Army Technical Manual TM 11-476, July 1947.
7. *ARRL Handbook,* American Radio Relay League, 1988, Chapter 39.
8. Jenkins, H.H., and R.W. Moss, "Error Reduction in Loop Direction-Finders," Final Report, Contract DA 28-043AMC-01207(E), Atlanta, GA: Georgia Institute of Technology, Engineering Experiment Station, September 1967.
9. Blakely, J.R., "U.S. Coast Guard Automatic Direction Finder Model RD132," *Trans. IRE,* Vol. CS-3, No. 1, March 1955, pp. 16–22.
10. Jay, F., ed., *IEEE Standard Dictionary of Electrical and Electronic Terms,* ANSI/IEEE Standard 100-1988, New York: Institute of Electrical and Electronic Engineers, Inc., 1988.
11. Giacolletto, L.J., and S. Stiber, "Medium-Frequency Crossed-Loop Radio Direction Finder with Instantaneous Unidirectional Visual Presentation," *Proc. IRE,* Vol. 37, No. 9, September 1949, pp. 1082–1088.
12. Watson-Watt, R.A., and J.F. Herd, "An Instantaneous Direct Reading Radio Goniometer," *J. IEE,* Vol. 64, May 1926, pp. 611–617.
13. Staff of the Radio Research Section, "A Short-Wave Cathode-Ray Direction-Finding Receiver," *Wireless Engineer and Experimental Wireless,* Vol. 15, 1938, p. 432.
14. Watson-Watt, R.A., "The Directional Recording of Atmospherics," *J. IEE,* Vol. 64, May 1926, pp. 234–238.
15. Ernst, E.W., "Digital Techniques for Radio Direction Finding," *Proc. Conf. HF Radio Propagation,* Urbana, IL: University of Illinois, August 1970, pp. 203–251.
16. Jenkins, H.H., R.W. Moss, and L. Scott, "Error Reduction in HF Loop DF Systems," *IEEE Trans. Aerospace and Electronic Systems,* Vol. AES-5, No. 5, May 1969, pp. 486–498.
17. Jenkins, H.H., and R.W. Moss, "Review of Small-Aperture DF Investigations," *Proc. Conf. HF Radio Propagation,* Urbana, IL: University of Illinois, August 1970, pp. 151–201.
18. Hedlund, D.A., and L.C. Edwards, "Polarization Fading over an Oblique Incidence Path," *IRE Trans. Antennas Propagat.,* Vol. AP-6, January 1958, pp. 21–25.
19. Terman, F.E., and J.M. Pettit, "A Proposal for the Reduction of Polarization Errors in Loop Direction Finders," *Proc. IRE,* Vol. 28, June 1940, p. 285.
20. Terman, F.E., and J.M. Pettit, "The Compensated-Loop Direction Finder," *Proc. IRE,* Vol. 33, May 1945, pp. 307–318.
21. Barfield, R.H., "The Performance and Limitations of the Compensated-Loop Direction Finder," *J. IEE,* Vol. 86, August 1940, pp. 396–398.
22. Jenkins, H.H., R.W. Moss, and L. Scott, "The Tilted Loop: An Error Reduction Technique for Loop Direction Finders," *IEEE Trans. Aerospace and Electronic Systems,* Vol. AES-5, No. 6, 1964.
23. Moore, J.D., and M.P. Castles, "HF Spaced Loop Antennas," Final Report, Contract DA 28-043 AMC-01960(E), San Antonio, TX: Southwest Research Institute, July 1967.
24. Moore, J.D., *et al.,* "The Design of the Modern Spaced-Loop DF Antenna for HF and VHF," *Proc. Fourth Allerton House Conf. Radiolocation Research,* Urbana, Illinois, June 1971, pp. 1.23–1.44.
25. Caplin, F., and J.H. Bagley, "A Mobile Spaced-Loop Direction-Finder," *J. IEE (London),* Vol. 94, Part IIIA, No. 15, 1947, pp. 676–682.
26. Moore, J.D., and M.P. Castles, "The Coaxial Spaced Loop as a Small-Aperture Multipolarization Radio Direction Finder Antenna," *Southwestern Conf. Record,* Dallas, Texas, April 1966, pp. 20–22.
27. Travers, D.N., "Characteristics of Electrically-Small, Spaced-Loop Antennas," *IEEE Trans. Antennas Propagat.,* Vol. AP-13, July 1965, pp. 639–641.

28. Horner, F., "An Experimental Spaced-Loop Direction-Finder for Very High Frequencies," *J. IEE,* Vol. 94, Part IIIA, 1947, pp. 126–133.

29. Ross, W., "The Development and Study of a Practical Spaced-Loop Radio Direction-Finder for High Frequencies," *J. IEE,* Vol. 94, Part III, 1947, pp. 99–125.

30. Pressey, B.G., "Rotating H-Adcock Direction-Finder," *Wireless Engineer,* March 1949, pp. 85–92.

31. Robertson, W.J., "A Low-Noise UHF Interferometer," Technical Report 2142-4, Columbus, OH: The Ohio State University Research Foundation, July 1966.

32. Priedigkeit, J.H., "The V-Scan Direction Finding Technique," *Proc. Fourth Allerton House Conf. Radiolocation Research,* Urbana, Illinois, June 1971, pp. 4.12–4.32.

33. Travers, D.N., "Spacing-Error Analysis of the Eight-Element Two-Phase Adcock Direction Finder," *Trans. IRE Antennas Propagat.,* Vol. AP-3, April 1955, pp. 63–65.

34. Travers, D.N., "The Effect of Mutual Impedance on the Spacing Error of an Eight-Element Adcock," *Trans. IRE Antennas Propagat.,* Vol. AP-4, January 1957, pp. 36–39.

35. Burtnyk, N., "Measurement of Polarization Errors of an Eight-Element and a Four-Element Adcock Antenna," Report ERB-503, National Research Council of Canada, March 1959.

36. Redgemont, P.G., "An Analysis of the Performance of Multi-Aerial Adcock Direction-Finding Systems," *J. IEE,* Vol. 94, Part IIIA, 1947, pp. 751–761.

37. Wait, J.R., and W.A. Pope, "Evaluation Errors in an Eight-Element Adcock Antenna," *Trans. IRE Antennas Propagat.,* Vol. AP-2, October 1954, pp. 159–162.

38. Hansel, P.C., "Instant-Reading Direction Finder," *Electronics,* April 1948, pp. 86–91.

39. Cleaver, R.F., "The Development of Single-Receiver Automatic Adcock Direction-Finders for Use in the Frequency Band 100–150 MHz," *J. IEE,* Vol. 94, Part IIIA, March 1947, pp. 783–797.

40. Berhert, J.H., "Effect of Tracking Accuracy Requirements on the Design of the Minitrack Satellite Tracking System," *Trans. IRE,* Vol. I-9, No. 2, September 1960, pp. 84–88.

41. Jenkins, H.H., and R.W. Moss, "Tiltable, Rotating Vertical Loop DF System," Annual Report, Contract DAAB07-68-C0072, Atlanta, GA: Electronics Division, Georgia Institute of Technology, September 1969.

42. Jenkins, H.H., *et al.,* "Improved Small-Aperture DF Systems," Special Report No. 1, Contract DAAB07-70-C0261, Atlanta, GA: Engineering Experiment Station, Georgia Institute of Technology, May 1971.

43. Lipsky, S.E., *Microwave Passive Direction-Finding,* New York: John Wiley and Sons, 1987, Chapter 9.

44. Bullock, L.G., G.R. Oeh, and J.J. Sparagna, "An Analysis of Wideband Microwave Monopulse Direction-Finding Techniques," *IEEE Trans. Aerospace and Electronic Systems,* Vol. AES-7, January 1971, pp. 188–202.

45. Tsui, J.B.Y., *Microwave Receivers and Related Components,* PB84-108711, Springfield, VA: National Technical Information Service, 1983, Section 7.8.

46. Wiley, R.G., *Electronic Intelligence: The Interception of Radar Signals,* Dedham, MA: Artech House, Inc., 1985, Section 4.4.

47. Harrison, C.W., "Antenna Coupling Errors in Direction Finders," *J. Res. NBS,* Vol. 65D, No. 4, July–August 1961.

48. Harrison, C.W., "Scattering Error in a Radio Interferometer," *Trans. IRE Antennas Propagat.,* Vol. AP-10, No. 3, May 1962, pp. 273–286.

49. King, R.W.P., *The Theory of Linear Antennas,* Cambridge, MA: Harvard University Press, 1956, Chapter 3.

50. Jordan, E.C., *Electromagnetic Waves and Radiating Systems,* Englewood, NJ: Prentice-Hall Inc., 1950, Chapter 15.

51. Jenkins, H.H., *et al.,* "Miniature Manpack DF Systems," Final Report, Contract DAAB07-71-

C0243, Atlanta, GA: Engineering Experiment Station, Georgia Institute of Technology, January 1973.

52. Ross, W., E.N. Bramley, and G.E. Ashwell, "A Phase Comparison Method of Measuring the Direction of Arrival of Ionospheric Radio Waves," *J. IEE,* Vol. III, No. 98, 1951, pp. 294–302.

53. Burtynyk, N., C.W. McLeish, and J. Wolfe, "Performance of an Interferometer Direction Finder for the HF Band," *Proc. IEE,* Vol. 112, No. 11, November 1965, pp. 2055–2059.

54. Leiner, B.M., "An Analysis and Comparison of Energy Direction Finding Systems," *IEEE Trans. Aerospace and Electronic Systems,* Vol. AES-15, No. 6, November 1979, pp. 861–873.

55. Lipsky, S.E., *Microwave Passive Direction-Finding,* New York: John Wiley and Sons, 1987, Chapter 9.

56. Bullock, L.G., G.R. Oeh, and N.J. Sparagna, "An Analysis of Wideband Microwave Monopulse Direction-Finding Techniques," *IEEE Trans. Aerospace and Electronic Systems,* Vol. AES-7, January 1971, pp. 188–202.

57. Torrieri, D.J., *Principles of Military Communications Systems,* Dedham, MA: Artech House, Inc., 1981.

58. Jacobs, E., and E.W. Ralston, "Ambiguity Resolution in Interferometry," *Trans. IEEE Aerospace and Electronic Systems,* Vol. AES-17, No. 6, November 1981.

59. Floyd, P., and J. Taylor, "Dual-Channel Space-Quadrature Interferometer System," *Microwave System Design Handbook,* 1987, pp. 133–148.

60. Johnson, R.L., *et al.,* "Short Time Scale Ionospheric Tilt Measurements for Lateral Deviation Compensation," *Proc. Fourth Allerton House Conf. Radiolocation Research,* Urbana, Illinois, June 1971.

61. Ernst, E.W., and K.D. Stengel, "An Interferometer DF System with On-Line Computers," *Proc. Fourth Allerton House Conf. Radiolocation Research,* Urbana, Illinois, June 1971.

62. Sherrill, W.M., *et al.,* "Advanced Radiolocation Interferometry," *Proc. Fourth Allerton House Conf. Radiolocation Research,* Urbana, Illinois, June 1971.

63. Bailey, A.D., and W.C. McClung, "A Sum-and-Difference Interferometer for HF Radio Direction-Finding," *IEEE Trans. ANE,* Vol. ANE-10, March 1963.

64. Treharne, R.F., *et al.,* "Some Characteristics of the Propagation of Skywaves over Short Ionospheric Paths," *Proc. IREE Australia,* 1965, pp. 245–254.

65. Fantoni, J.A., and R.C. Benoit, "Applying the Doppler Effect to DF Design," *Electronic Industries,* Vol. 16, No. 1, January 1957, and Vol. 16, No. 2, February 1957.

66. Earp, C.W., and D.L. Copper, "The Practical Evolution of the Commutated-Aerial Direction-Finding System," *Proc. IEE,* Paper 2569R, Vol. 105, Part B, Supplement No. 9, March 1958, pp. 317–332.

67. Oden, M., "Rate the Merits of Pseudo-Doppler Direction-Finding," *Microwave and RF,* March 1989, pp. 79–81.

68. Fantoni, J.A., and R.C. Benoit, "Doppler-Type High Frequency Direction Finder," *Conf. Record IRE,* March 1956, pp. 165–171.

69. Rogers, T., "A Doppler ScAnt," *QST,* May 1978, pp. 24–28.

70. Skolnik, M.I., *Introduction to Radar Systems,* New York: McGraw-Hill Book Co., 1962, Chapter 10.

71. Oetzel, G.N., and E.T. Pierce, "VHF Technique for Locating Lightning," *Radio Sci.,* Vol. 4, No. 3, March 1969, pp. 199–202.

72. Chanos, N., G.N. Oetzel, and E.T. Pierce, "A Technique for Accurately Locating Lightning at Close Range," *J. Appl. Meteorol.,* Vol. 11, October 1972, pp. 1120–1127.

73. Torrieri, D.J., "Adaptive Thresholding System," *IEEE Trans. Aerospace and Electronic Systems,* Vol. AES-13, May 1977, p. 273.

Chapter 6
REPRESENTATIVE OPERATIONAL
SMALL-APERTURE
DIRECTION-FINDING SYSTEMS

The physical implementations of small-aperture DF systems manifest many diverse forms due to the wide variety of DF techniques and applications. This chapter presents and discusses representative hardware implementations for the amplitude, differential phase-to-amplitude, and phase DF techniques.

6.1 AMPLITUDE DIRECTION-FINDING SYSTEMS

Small-aperture amplitude-response DF systems are extensively used in the MF and HF bands for aircraft and marine RDFs and ADFs.

6.1.1 Aircraft MF and HF ADF Systems

Aircraft ADF systems operate in the 190- to 1799-kHz region and provide azimuthal bearing information by using transmissions from the established aeronautical radio–navigation ground beacon network and AM broadcast stations. The signal sources are at known locations and provide inherent identification. The sources use carrier-based transmissions that are continuous, narrowband, and employ modulation formats that are highly compatible with DF operation. Military aircraft ADF systems may use specially deployed transportable radio ground beacons operating in the 200- to 535-kHz and 1650- to 1750-kHz frequency ranges.

Aircraft ADFs employ orthogonal crossed-loop antennas that are electronically scanned using the technique depicted in Figure 5.8. One loop is positioned parallel to the centerline of the fuselage and provides a $\sin\phi$ response where ϕ is the azimuth angle of the signal source relative to the nose of the aircraft. The loop perpendicular to the aircraft centerline produces a $\cos\phi$ response. Allen and Ryan

[1] discuss the design of ADF aircraft loop antennas, techniques for installing the loops on the aircraft, and typical performance results. Emphasis is placed on the distortion created by the airframe in the *H*-field (magnetic) component of the incident signal and the effects on bearing error.

Figure 6.1 is a general block diagram that is typical of 1990 ADF systems. Maximum use is made of digital processing. Waveforms at points A through H on Figure 6.1 are shown in Figure 6.2 based on a signal arriving at an azimuth angle of 45°. In Figure 6.2, arrows pointing up and down represent phases leading and lagging the sense antenna phase.

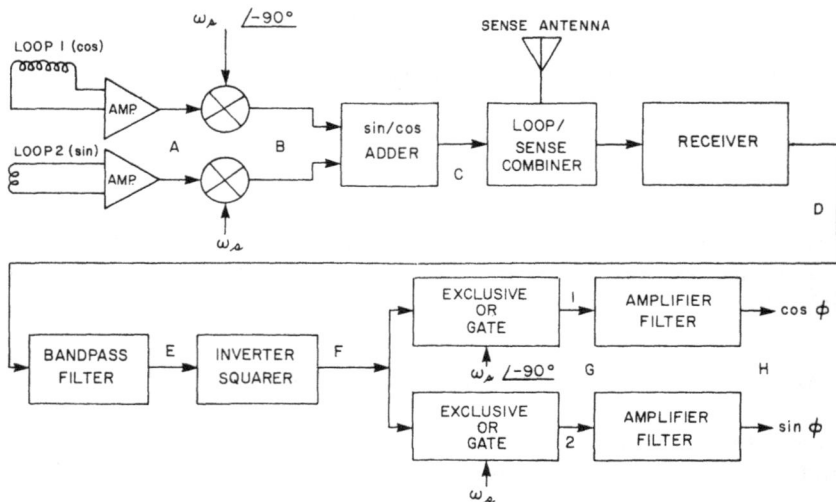

Figure 6.1 General block diagram of an aircraft ADF system.

Referring to Figure 6.2, after amplification, each loop signal is applied to a balanced modulator, which is switched by square waves. Figure 6.2 illustrates the balanced modulator digital switching waveform ω_s and $\omega_s/-90°$. A typical switching rate is 31 Hz. The balanced modulator reverses the phase of a loop signal when the switching signal is negative; therefore, the balanced modulators (B) reverse phase at a rate of ω_s. When the modulated loop outputs are combined (C), the phase reverses at a rate of $2\omega_s$. The sum of the sine and cosine channel signals is added to the sense antenna output to create an RF signal that is phase modulated. The RF phase leads or lags the RF sense signal if the combined loop signal phase is $-90°$ or $+90°$, respectively. The phase modulation of the combined RF signal is continuously variable in phase, where the amount of phase shift is a function of combined

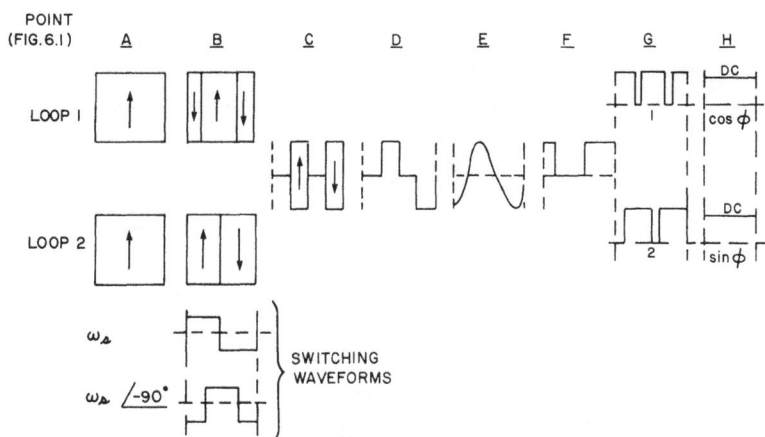

Figure 6.2 Waveforms for the ADF system of Figure 6.1.

loop signal amplitude. Therefore, the azimuth angle of arrival is phase encoded on the combined sense-loop RF signal.

After reception and detection, a phase-modulated baseband, stepped waveform is produced (D). This stepped waveform is bandpass filtered at a center frequency of ω_s. The resultant sinusoidal waveform (E) is converted to a unipolar square wave (F), which is synchronously detected with the switching waveforms ω_s and $\omega_s/-90°$ in exclusive-or gates. After amplification and filtering, the outputs at H are dc voltage levels representing $\cos\phi$ and $\sin\phi$ of the azimuth angle of arrival. The $\sin\phi$ and $\cos\phi$ measures of azimuth angle are used to drive both analog and digital bearing indicators. Generally, the analog display uses a resolver with a needle indicating bearing on a 360° compass card. Often the ADF bearing display is combined with a magnetic compass in a radio magnetic indicator (RMI).

ADFs function as a primary asset for both commercial and general aviation application. Three exemplary systems, developed by Collins Avionics, are discussed in the following sections.

The Collins ADF-650A, shown in Figure 6.3, operates from 200 to 1799 kHz in 1-kHz steps and is designed especially for general aviation use. Figure 6.4 shows the Collins ADF-60 Pro Line ADF, which operates from 190 to 1749.5 kHz in 0.5-kHz increments. The Collins ADF-462 Pro Line II system adds the 2.182-MHz maritime distress frequency to its coverage. The ADF-462 uses all-digital techniques and includes self-diagnostic capability. The Pro Line II supports the avionics common serial data bus (CSDB) and ARINC 429 data protocols. Both the ADF-60 and ADF-462 accommodate either single-loop or dual crossed-loop antennas.

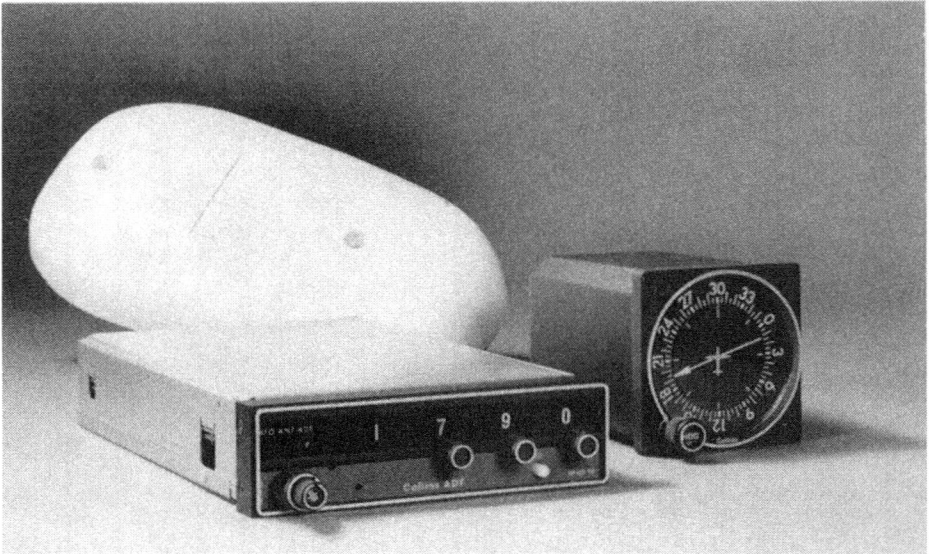

Figure 6.3 ADF system suitable for general aviation use. (Courtesy of Rockwell International.)

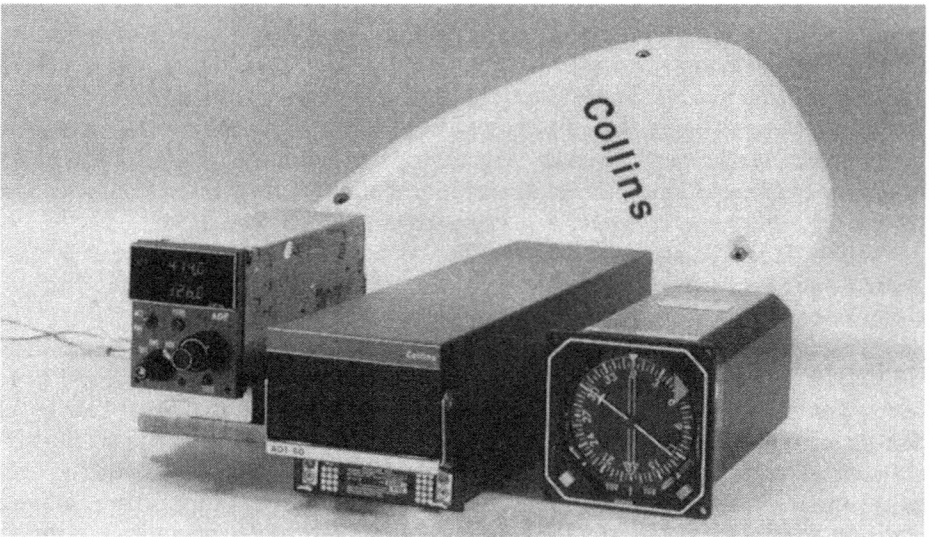

Figure 6.4 ADF system suitable for commercial and business aircraft use. (Courtesy of Rockwell International.)

The ADF-60, 462, and 650A use microelectronic technology to reduce size, weight, and power consumption. Loop and sense antenna functions are combined into a single antenna unit. A built-in sense antenna RF preamplifier eliminates the need for a long-wire sense antenna. Loop and sense RF preamplifiers preclude the need for special critical length RF cabling between the antennas and receiver. All frequencies are digitally derived from a single-crystal synthesizer. A synchronous filter improves performance under heavy atmospheric noise conditions.

Typical bearing accuracy for aircraft ADF systems, using antennas designed to the ARINC 712 specification, is $\pm 2°$ rms for an incident field strength of 20 μV/m. Accuracy increases to about $\pm 1°$ rms for a 35 μV/m field strength.

6.1.2   Marine MF and HF RDF–ADF Systems

Marine DF systems using direct amplitude response include both RDF and ADF units. The RDF units use single-loop and sense antennas and manual rotation for amplitude null sensing. The RDF manually scanned, battery-operated, single-loop systems generally use a ferrite antenna with a basic superheterodyne receiver operating in the marine and aircraft beacon band of 180 to 410 kHz and the AM commercial broadcast band of 550 to 1600 kHz. Crystal-controlled channels are required for instant frequency tuning, and a beat frequency oscillator is a necessity for SSB reception. Both fixed, surface-mounted and hand-held versions exist. The hand-held RDF, which generally covers only the beacon band, is especially useful for homing. Further, the hand-held RDF can be positioned on a boat where radio deviation effects (bearing errors) are at a minimum. Special-purpose marine single-loop RDFs operate in the 26- to 32-MHz band on radiobeacon-marked waterborne objects and buoyed locations. A frequency of 27 MHz is widely used for small cetacean radiotracking using hand-held RDFs [2]. The VHF frequency band from 148 to 172 MHz is also used for radiotracking of small cetaceans [2]. (Biotelemetry is often combined with the radiotracking of wildlife [3].)

The ADF systems use crossed-loop antennas that are either electronically scanned or electromechanically scanned by a goniometer. The electronic-scanning units are virtually identical to the aircraft ADF versions discussed in Section 6.1.1. The electronic-scanning units are typically used on the smaller boats and cover only the beacon, AM broadcast, and distress frequencies. The crossed-loop, goniometer-scanned MF and HF systems are primarily used on larger vessels and shore stations for use in general navigation and rescue operations. These systems provide DF information over the 100-kHz to 30-MHz frequency range.

Figure 6.5 shows a system that operates from 100 kHz to 30 MHz in 100-Hz steps with a DF mode operating to 18 MHz. The antenna is an 85-cm-diam crossed loop with a 2-m-long sense antenna. Sense is automatic. The receiver is a synthesized, triple-conversion superheterodyne. Two hundred channels may be preset

Figure 6.5 Example of a marine MF and HF ADF system. (Courtesy of Furuno Electric Co.)

into memory. Typical bearing accuracy is $\pm 1°$ rms at a 1 μV/m field strength. Reception modes are A1A, A2A, H2A, A3E, H3E, and J3E. (The LED digital display shows operating frequency.)

Figure 6.6 is a crossed-loop, goniometer-scanned ADF operating over the 200-kHz to 18-MHz band in 100-Hz increments, with 400-channel frequency memory. The DF mode operates to 16 MHz. The unit can search over a 500-kHz range about a selected center frequency in 2-kHz steps. The receiver is a synthesized, triple-conversion superheterodyne. The antenna is an 80-cm-diam crossed loop with inherent sense capability that requires no external sense antenna. Typical bearing accuracy is $\pm 1°$ rms at 100 μV/m incident field strength. Reception modes are A1A, A2A, A3E, H2A, H3E, and J3E. The LED display provides channel, frequency, and signal strength information.

An automatic digital direction finder (ADDF) is shown in Figure 6.7. Using a crossed-loop antenna and operating from 0.190 to 4.5 MHz, the unit provides both digital and analog bearing read-outs. The receiver is a single-conversion superheterodyne capable of operating on A1A, A2A, A3E, and J3E modulations. The minimum measurable field strength is 8 μV/m in the beacon band, 7 μV/m in the AM broadcast band, and 5 μV/m in the marine band.

The accuracy of marine DF systems is a function of operating frequency and type of emitter. The best accuracy is obtained within the "service range" of marine

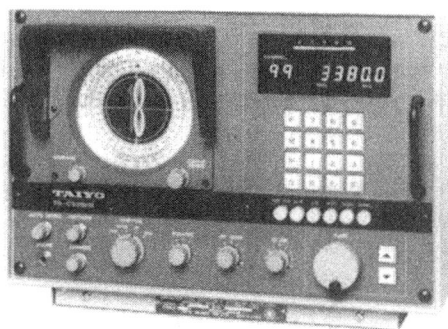

Figure 6.6 Example of a marine MF and HF ADF system. (Courtesy of Taiyo Musen Co.)

radiobeacons operated by the U.S. Coast Guard (USCG) on frequencies between 285 and 325 kHz. The rated service range, locations, and other detailed characteristics of marine radiobeacons are given in the USCG Light List available from the Superintendent of Documents, Government Printing Office. Attempts to operate beyond the rated service range of a marine radiobeacon may result in degraded performance due to a combination of reduced SNR and sky wave interference.

Figure 6.7 Example of a marine MF and HF ADF system. (Courtesy of Taiyo Musen Co.)

6.2 DIFFERENTIAL PHASE-TO-AMPLITUDE DIRECTION-FINDING SYSTEMS

Small-aperture DF systems using differential phase-to-amplitude techniques are widely employed for marine, government, and public service VHF ADF applications. Portable phase-to-amplitude DF systems are used for location and homing-on VHF distress emitters such as the marine emergency position-indicating radio-beacon (EPIRB) and emergency locator transmitter (ELT). Portable, mobile, and fixed phase-to-amplitude DF systems are integral elements in tactical intelligence collection systems. The "instantaneous" Watson-Watt DF system is proving to be very effective on both uncooperative and cooperative frequency-hopping spread-spectrum transmissions. The following sections discuss representative operational phase-to-amplitude DF systems.

6.2.1 Marine, Government, and Public Service VHF DF Systems

The most common implementation is based on dual-baseline electronically scanned Adcock arrays. Figure 6.8 shows a typical mast-mounted antenna config-uration with the balanced modulators for electronic scanning installed in the antenna unit assembly. Figures 6.9 through 6.11 illustrate representative marine and weather band ADF units. (The 121.5-MHz distress frequency is also covered.) Synthesized, double-conversion, superheterodyne receivers are used. (The system shown in Figure 6.9 uses a separate VHF marine transceiver as the receiver.) Typ-ical sensitivity is 1 μV/m or less for DF operation, and a nominal DF accuracy is $\pm 1°$ rms. The operational accuracy is determined by installation and site factors. Bearing indication response time is equal to or less than 0.1 seconds. Modulation capability is F3E (A3E on 121.5-MHz EPIRB transmissions.)

Figure 6.8 Orthogonal Adcock array for marine ADF use. (Courtesy of Apelco Marine Electronics.)

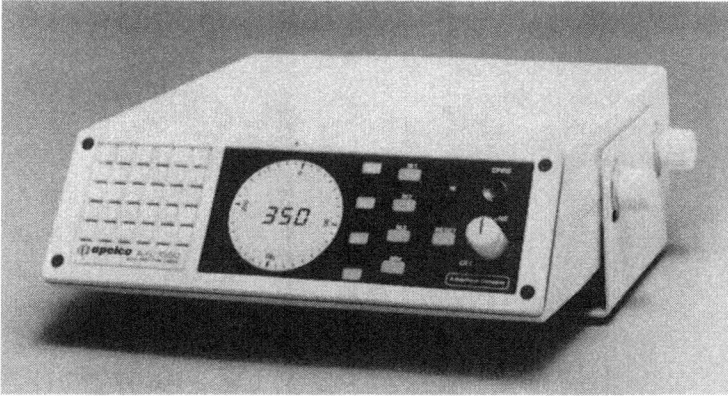

Figure 6.9 Example of a marine VHF ADF system. (Courtesy of Apelco Marine Electronics.)

The systems shown in Figures 6.10 and 6.11 provide both digital and direct bearing read-outs. Direct bearing data are indicated by LEDs positioned around the circular compass card.

Figure 6.10 Example of a marine VHF ADF system. (Courtesy of Taiyo Musen Co.)

Figure 6.11 Example of a marine VHF ADF system. (Courtesy of Taiyo Musen Co.)

An ADF that covers the 110- to 170-MHz government, marine, and public service band is shown in Figure 6.12. A synthesized, triple-conversion superheterodyne receiver provides a rated DF sensitivity of 0.5 μV/m. Modulation capability is A2A, A3E, and F3E.

Figure 6.12 Example of a marine VHF ADF system. (Courtesy of Taiyo Musen Co.)

6.2.2 Emergency and Distress VHF

Dire emergency and distress VHF transmissions are provided by ELTs in the aircraft band and EPIRBs in the marine band. These beacons operate on 121.5 MHz.

Transmissions of 406 MHz may also be used for satellite detection and doppler location. Homing is performed on the 121.5-MHz source after location in a general area by the satellite detection and location. Many marine and aircraft DF systems operate on 121.5 MHz. An example is depicted in Figure 6.13. This portable DF system is especially designed for both airborne and surface-based operation on 121.5 MHz plus any other three channels between 118 and 136 MHz. Spaced dipoles operating in the phase differential-to-amplitude mode are the most commonly used antenna for the system shown in Figure 6.13.

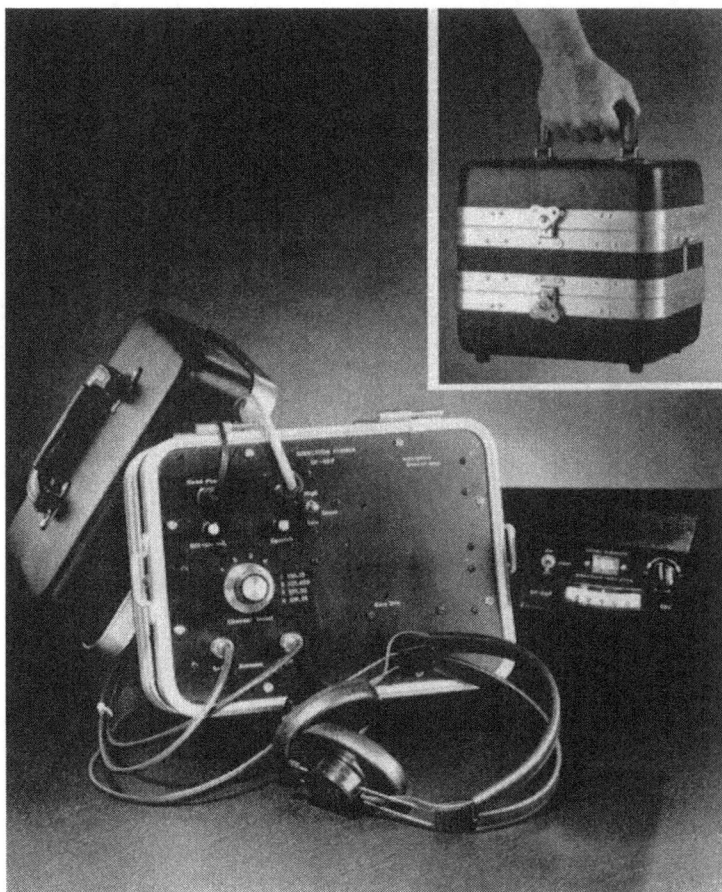

Figure 6.13 A portable system for distress and emergency frequency DF. (Courtesy of Emergency Beacon Corp.)

6.2.3 Tactical Intelligence Collection Systems

Tactical intelligence collection systems perform HF, VHF, and UHF intercept and DF functions in transportable and manpack configurations. An exemplary system is shown in Figures 6.14 and 6.15. The system is the Watkins-Johnson Co. WJ-8990 manpack tactical intelligence system (MANTIS). Figure 6.14 is the WJ-9081 antenna; Figure 6.15 is the WJ-8972 receiver-DF processor unit. The basic MANTIS system provides intercept and DF from 20 to 500 MHz, which is expandable from 0.5 to 1100 MHz for intercept and 20 to 1100 MHz for DF.

Figure 6.14 Antenna system for a tactical intelligence collection DF. (Courtesy of Watkins-Johnson Co.)

Figure 6.15 Receiver and DF processor for a tactical intelligence collection DF. (Courtesy of Watkins-Johnson Co.)

The WJ-9881 DF antenna is a two-bay unit with each bay using an orthogonal Adcock array. The upper bay operates from 140 to 500 MHz with a baseline spacing of 0.36λ at 500 MHz. The upper bay vertical dipole antennas, which are λ/2 at 500 MHz, are resistively loaded with an integral preamplifier to provide broadbanding and enhanced DF accuracy. The vertical dipole lengths are a half-wavelength at 500 MHz. The lower bay operates from 20 to 140 MHz with a baseline spacing of 0.498λ at 140 MHz. The vertical dipoles, which are less than λ/2 at 140 MHz, use active antenna technology to improve performance at the lower frequencies while reducing the physical size. The antenna bays are mounted on a lightweight mast, composed of carbon fiber material, which can be erected to 40 feet by one or two persons in less than 15 minutes.

The receiver-DF processor unit (WJ-8990) has a DF frequency range of 20 to 500 MHz extendable to 20 to 1200 MHz. This unit can operate either as a standalone DF system or as an element of a DF net to provide emitter position location. In the netted mode, the unit can serve as the net control station (NCS) and provide tasking to one to three outstations via an external RS-232 data link. As a net control station the unit can interrogate remote outstations and receive LOB data from each outstation to compute a position-fix on the subject emitter. Bearing accuracy is rated at 3° rms. In the netted mode, an emitter position can be fixed to UTM coordinates at a 50% CEP radius in 10 seconds.

6.3 PHASE DIRECTION-FINDING SYSTEMS

DF systems based on phase measurements include two major types: phase interferometers and pseudo doppler. Representative operational versions of these two types are discussed in the following sections.

6.3.1 Interferometers

Direct phase comparison interferometer DF systems are well-suited for tactical intelligence collection and electronic support measures (ESM). Both require near-instantaneous DF over a broad frequency range with precise angle-of-arrival measurement accuracy. Further, the signals of interest include a variety of modulation formats. The direct phase DF is relatively tolerant of signal modulation including short-duration and spread-spectrum transmissions.

The current technology of phase DF for tactical intelligence collection and ESM applications is rapidly improving. An exemplary system is the Watkins-Johnson Co. WJ-8976 three-channel DF system [4], which operates from 20 to 500 MHz with extendable coverage down to 2 MHz and up to 1200 MHz. The WJ-8976 system is based on the triple-baseline, triple-channel phase interferometer principle diagrammed in Figure 5.53. The unit employs antenna systems consisting of three electrically short vertical dipoles arrayed in an equilateral triangle with the baselines less than a half-wavelength at the highest operating frequency.

Figure 6.16 illustrates the WJ-8976 antenna configurations for full 2- to 1200-MHz operation. The HF antenna, which covers 2 to 30 MHz, consists of three vertical center-fed dipoles designed to be mounted directly on the ground. The dipole lengths are 3.96 m or 0.396λ at 30 MHz. The baselines are 4.1 m or 0.41λ at 30 MHz. The VHF and UHF arrays are mast-mounted and configured into three separate bays covering 20 to 100 MHz (lower bay), 100 to 500 MHz (mid-bay), and 500 to 1200 MHz (high bay). Dipole lengths and baselines are all less than one-half wavelength. Each antenna element drives a low-noise preamplifier centrally located at the configuration feedpoint. RF calibration signals are inserted between the antenna elements and the associated preamplifiers.

A system-level, functional block diagram of the WJ-8976 system is shown in Figure 6.17. The basic subsystems are one or more antenna units, a triple-channel slave receiver operating in conjunction with a master surveillance receiver and a digital DF processor. Figure 6.18 shows the antenna bay selector (top), DF processor (middle), and the 20- to 500-MHz slave receiver (bottom). All local control is provided by the DF processor using the IEEE-488 bus protocol. Remote control capability is included to provide for (1) netted mode operation with remote outstations and (2) interfacing with external equipment, such as tactical computers, for capability expansion. The DF processor bearing display is a circular array of LEDs

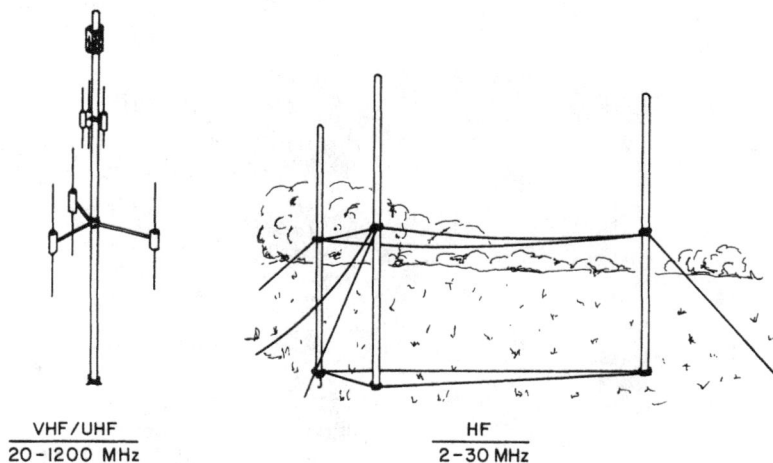

Figure 6.16 Phase interferometer antennas for a tactical DF system. (Courtesy of Watkins-Johnson Co.)

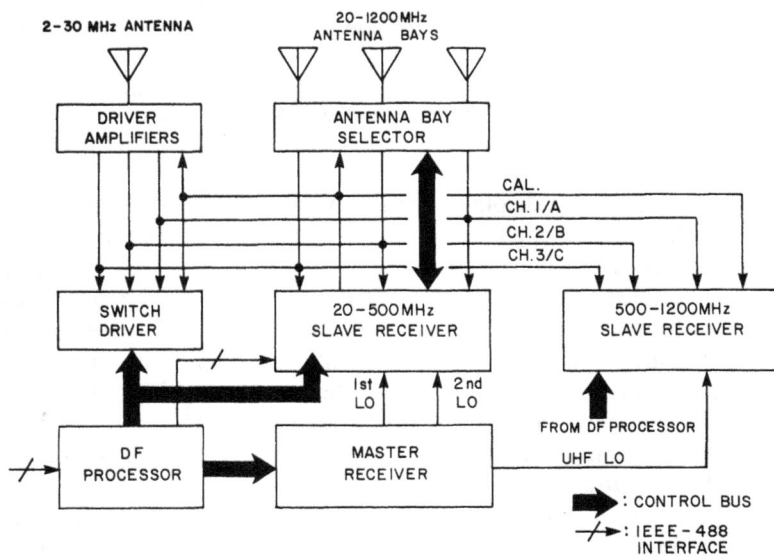

Figure 6.17 Functional block diagram of a HF and VHF phase interferometer DF. (After Hedges [4], Figure 1.)

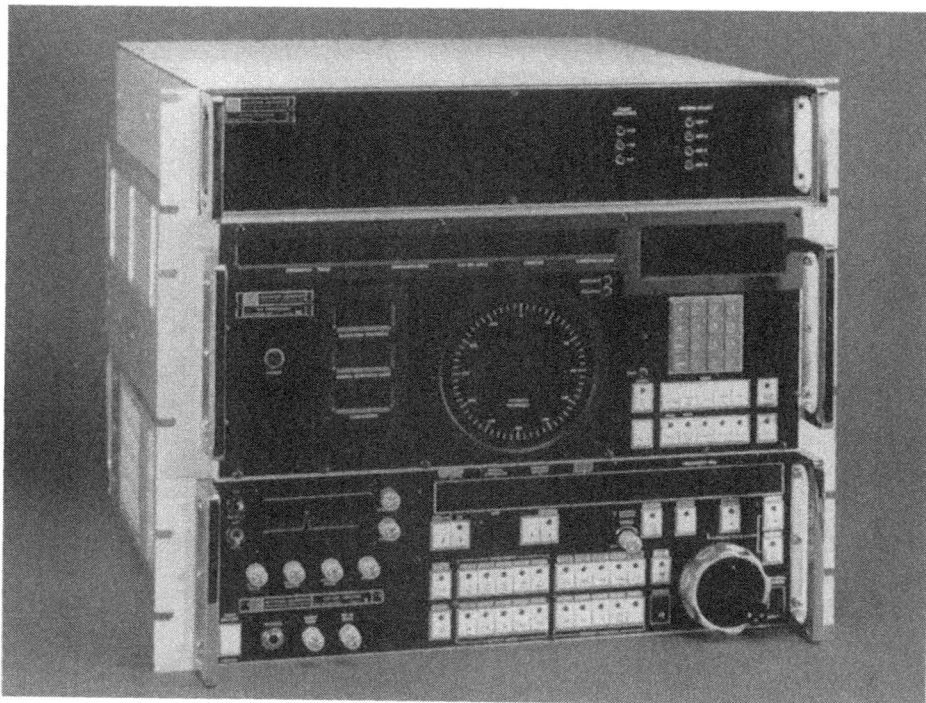

Figure 6.18 HF and VHF phase interferometer showing the antenna bay selector (top), DF processor (middle), and slave receiver (bottom). (Courtesy of Watkins-Johnson Co.)

surrounding a 360° compass display. The illuminated LEDs occur at the azimuth angle of arrival. An integral digital read-out also displays the bearing angle. Additional digital displays are tuned frequency, IF bandwidth, resolution bandwidth, signal strength, signal threshold, elevation angle, and AOA confidence factor.

The RF outputs of the selected antenna bay are continuously connected to the triple-channel inputs of the slave receiver, as diagrammed in Figure 6.19. The first and second synthesized LOs are from the master receiver. This technique ensures that all three slave receiver channels are coherently tuned and tracking. After quadruple-conversion, the slave receiver outputs are three 500-kHz signals with 100-kHz bandwidths.

The RF calibration signal is generated in the slave receiver by up-converting a 21.4-MHz reference generator. The RF calibration signal is periodically inserted at the antenna preamplifier inputs to measure, and then remove, gain and phase mismatches between the three receiver channels. This practice removes the need for identically matched channels.

The DF processor is diagrammed in Figure 6.20. The 500-kHz IF inputs are sampled at a 400-kHz rate and synchronously digitized to 12 bits of resolution. (The Nyquist sampling rate criteria may be exceeded because of the bandpass nature of the sampled signals and the subsequent application of discrete frequency transform algorithms. No aliasing occurs.) The digitized data for each channel are blocked into frames of up to 1024 samples. The duration of the data frames may be operator-adjusted between 10 μs and 5.12 ms. The longer the data samples, the more accurate the estimated angle of arrival. However, the DF accuracy on short-duration signals is enhanced when the durations of the signal and data frames coincide. The framing may be either programmed or triggered. An external trigger is usually obtained from the carrier operated relay (COR) output of the master receiver. This ensures that data framing is initiated only by the presence of a reliable signal. Using the external trigger mode, signals as short as 10 μs may be reliably sampled. The internal trigger is obtained from the slave receiver and occurs when the slave receiver's IF outputs exceed an internal threshold level in the DF processor. The threshold level is set above the receiver noise floor.

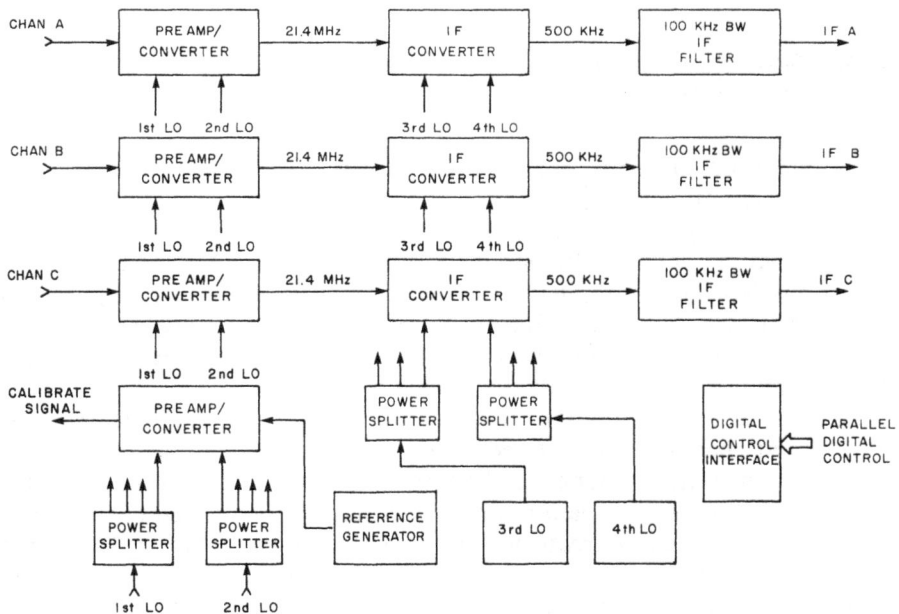

Figure 6.19 Block diagram of a phase interferometer receiver. (From Hedges [4], Figure 2. Reprinted with permission.)

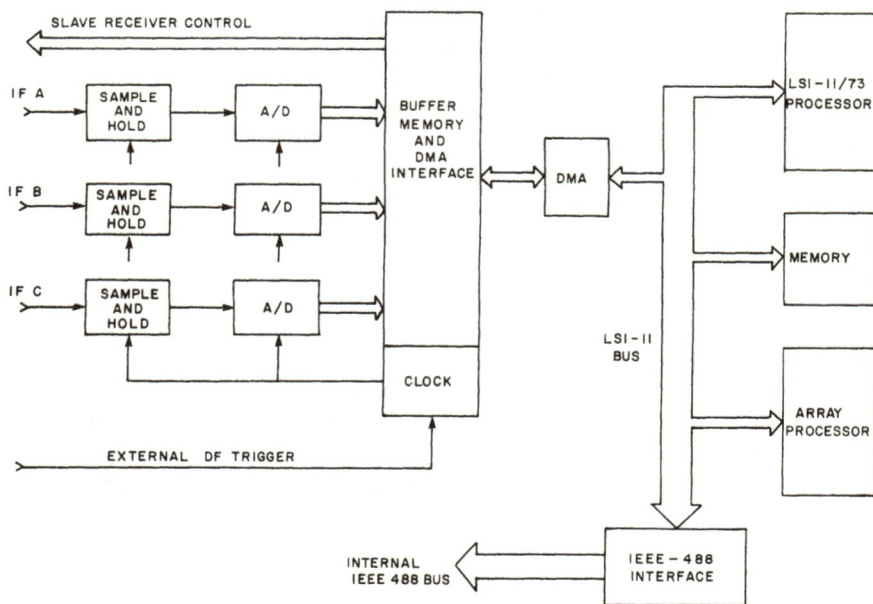

Figure 6.20 Block diagram of a phase interferometer DF processor. (From Hedges [4], Figure 3. Reprinted with permission.)

In Figure 6.20, the digitized IFs are loaded into the buffer memory by high-speed direct memory access (DMA) circuits. Computations are performed by the LSI-11/73 minicomputer operating in conjunction with a high-speed floating point processor. Two major computations are performed. First, a discrete Fourier transform (DFT) algorithm is applied to the digitized data to extract amplitude and phase parameters as discussed in Section 5.4.3. Second, the DFT data are used, with the geometry of the antenna and the operating frequency, to determine angle-of-arrival information.

Angle-of-arrival estimation involves correlating the DFT-derived signal amplitude and phase data with stored digital data sets representing gain and phase response data measured under controlled test conditions. (Chapter 9 discusses controlled testing for phase and gain response calibration.) The stored digital data sets are defined as a function of frequency and elevation and azimuth angles of arrival. The stored and measured complex digital data sets are correlated, and an initial measure of the angle of arrival is assigned to the angle associated with the stored data that maximizes the magnitude of the complex correlation coefficient. Correlation is performed at AOA increments of 10 to 12 degrees. This incrementation accommodates the conflicting requirements of computation response time and data memory storage.

Figure 6.21 is an example of azimuth angle correlation coefficient values at 12-degree intervals in azimuth; the true azimuth angle of the signal is 175 degrees. The estimated angle of arrival is determined by parabolic curve-fitting in the vicinity of the maximum correlation coefficient. A parabolic curve is fitted to the maximum correlation coefficient value and the two adjacent values. The estimated azimuth angle is the azimuth point at which the parabola achieves its maximum value. The minimum signal duration for an estimated angle-of-arrival measure is 10 μs. (This time does not include LO settling time. The IF outputs settle to within 1% of their final amplitude and phase values 35 μs after LO application.)

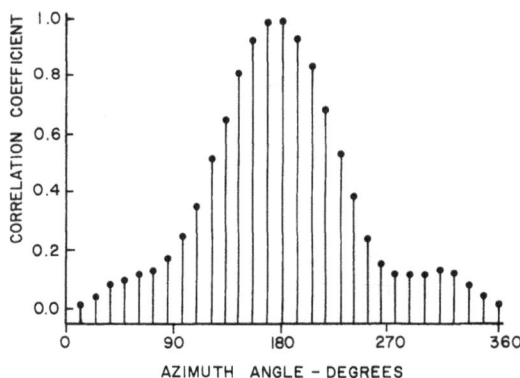

Figure 6.21 Example of azimuth angle correlation coefficients. (From Hedges [4], Figure 4. Reprinted with permission.)

Angle-of-arrival measurement accuracy is 2° rms based on standard Watkins-Johnson acceptance test procedures. Resolution is 0.01° for azimuth angle measures and 0.1° for elevation angle measures. Under operational conditions, the azimuthal error is given by

$$\phi_\epsilon = (94) \frac{\sqrt{B_p/N_a}}{X_t F_s d} \tag{6.1}$$

where

    ϕ_ϵ = the azimuthal error in degrees;
    B_p = the processor bandwidth in kHz;
    X_t = the antenna transfer function;
    F_s = the field strength in microvolts per meter;

d = the antenna baseline in meters;
f = the operating frequency in MHz;
N_a = the number of data frames averaged.

The antenna transfer function X_t is computed from

$$X_t = \frac{\text{antenna output voltage into } 50\ \Omega}{\text{incident field strength (V/m)}} \qquad (6.2)$$

The W-8976 system may be expanded by adding a graphic workstation. Graphical presentation of DF data greatly enhances an operator's ability to evaluate and assess the data and to recognize trends and patterns in the data. A combination of both rectangular and polar display formats is very effective. Especially effective are real-time displays of parametric functions such as azimuth *versus* time in a rectangular plot and signal strength *versus* azimuth in a polar plot.

6.3.2 Pseudodoppler

Until recently, most pseudodoppler DF systems operated from stationary or semi-mobile sites on friendly, cooperative transmissions. Typical applications were ship and aircraft identification using conventional VHF radio communication in the marine and aircraft frequency bands. The advent of microelectronic technology, efficient electrically small antennas, and digital processing has resulted in mobile, tactical pseudodoppler DF system implementations for use on uncooperative, hostile signals.

A representative tactical, portable, pseudodoppler DF is the Rohde and Schwarz GmbH & Co. PA 1100 system operating from 20 to 1000 MHz, which is covered using two antenna systems in the 20- to 200-MHz and 200- to 1000-MHz bands. Each antenna system consists of six circularly disposed vertical antennas and a center-mounted monopole. The antennas are scanned at a 340-Hz rate with alternating clockwise and counterclockwise scanning to compensate for signal doppler and modulation-dependent frequency shift. The PA 1100 also uses a proprietary frequency compensation technique [5] that affords single-channel operation rather than the dual-channel compensation discussed in Section 5.4.8.

The PA 1100 bearing information is displayed on an analog LCD compass rose with 5° resolution and a three-digit LCD digital read-out of bearing. The system bearing error limit is rated 2° rms for field strengths much greater than the minimum discernible field strength level. For ±5° bearing error, the minimum field strength level, using the 20- to 200-MHz antenna, is cited as 10 μV/m at 20 MHz decreasing to less than 3 μV/m between 40 and 200 MHz. With the 20- to 100-MHz antenna, the field strength for a ±5° bearing error is rated at a nominal value of 0.5 μV/m.

The PA 1100 contains a V.24/RS-232-C data interface capability for DF netting or remote control operation. The Rohde and Schwarz PA 010 HF DF pseudodoppler system covers 1 to 30 MHz using a fixed-site antenna system consisting of 32 crossed-loop antennas configured in two concentric circles. The large-diameter array covers 1 to 10 MHz, and the smaller diameter array covers 10 to 30 MHz. A center-located crossed-loop element acts as a reference for frequency compensation and signal surveillance and intercept. Rated azimuth accuracy is 1° rms. Alternating clockwise and counterclockwise antenna scanning is used to compensate for DF phase errors created by group-delay effects. The PA 010 includes an elevation angle-of-arrival measurement capability, which is augmented by single station location computerized software and hardware.

REFERENCES

1. Allen, W.P., and C.E. Ryan, Jr., "Aircraft Antennas," Chapter 37 in *Antenna Engineering Handbook,* Johnson, R.C., and H. Jasik, eds., New York: McGraw-Hill Book Co., 1984, pp. 37-1–37-14.
2. Scott, M.D., A.B. Irvine, R.S. Wells, and B.R. Mate, "Tagging and Marking Studies on Small Cetaceans," Chapter 28 in *The Bottlenose Dolphin,* Leatherwood, S., and R.R. Reeves, eds., New York: Academic Press, 1990, pp. 489–514.
3. Amlaner, C.J., and D.W. MacDonald, eds., *A Handbook on Biotelemetry and Radiotracking,* New York: Pergamon Press, 1990.
4. Hedges, S.A., "Three Channel Direction Finding System," Technical Note (unnumbered), Watkins-Johnson Co., March 1987.
5. Schnengler, Eckard, "Compensation of Modulation-Dependent Frequency Fluctuations of Single-Channel Radio Direction Finders," Federal Republic of Germany Patent DE 36 36 630 C1, published 21 April 1988. Described in the "News from Rohde and Schwarz," Vol. 30, No. 129, 1990/II, p. 42.

Chapter 7
PASSIVE GEOLOCATION

7.1 BASIC TECHNIQUES

Direction-of-arrival information obtained from small-aperture DF systems is used for many purposes; however, the primary goals are (1) passive geolocation of a subject emitter at an unknown location and (2) passive geolocation of the DF system relative to an emitter. Five basic techniques are used to accomplish these goals.

1. *Emitter homing:* DOA information on the subject emission is used by the DF system to move the DF system physically to the position of the emitter, which may be at a known or unknown location.
2. *Navigation:* DOA information on the subject emission originating from a known location is used by a moving DF system for navigation along a desired path. The emitter location is not the desired end point of the DF system path; the emitter provides way-point reference bearings only.
3. *Resection (back) triangulation:* DOA information obtained on multiple emitters at known locations is used at the DF site to geolocate the DF site, which is at an unknown location.
4. *Horizontal (azimuth) triangulation:* The geolocation of an emitter is determined by using azimuth bearing information from either multiple, dispersed DF systems or a single DF system that is properly moved relative to the subject emitter.
5. *Vertical triangulation:* Vertical triangulation obtains HF emitter location by measuring the azimuth and elevation angle of a sky wave signal arriving at the DF site. Knowing the height of the ionosphere at the refraction point, the measured elevation angle can be used in ray-tracing algorithms to determine the distance to the emitter. Emitter geolocation is computed from the azimuth and distance data.

Airborne vertical triangulation, called AZ/EL DF, is performed using measured azimuth and depression (elevation) angles and their intersection with the earth's surface.

The following sections discuss the five major passive geolocation techniques.

7.2 EMITTER HOMING

A DF system with no angle-of-arrival error may use DOA information to home-on an RF emitter by the most direct and shortest path. If the DF system experiences random DOA errors, the homing path will be erratic but it will eventually converge on the emitter. If a systematic bearing error exists, the homing path to the emitter will be a logarithmic spiral path, assuming that the DF system is consistently moved along the indicated bearing to the emitter [1]. The plot of a logarithmic spiral homing pattern in the xy-plane is shown in Figure 7.1. The DF system moves along the log spiral path to the emitter. The systematic error is the angle between the tangent to the homing path and the direct path to the emitter. (In Figure 7.1, the systematic bearing error Δ of about 45° is large to better illustrate the homing process.) The systematic bearing error remains constant along the log spiral homing path. Techniques have been developed to exploit this constant angle condition and estimate the emitter location without physically moving the DF system to the emitter. These techniques are based on horizontal triangulation, which is discussed in Section 7.5.

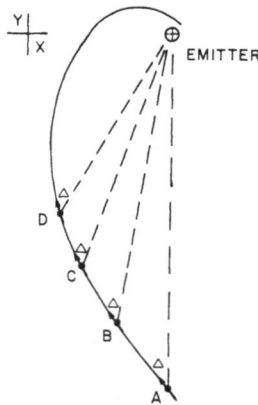

Figure 7.1 Logarithmic spiral homing trajectory.

In practice, ideal homing patterns, such as the log spiral, are distorted, especially in built-up areas and irregular terrain. Experience has shown that efficient homing is operator dependent and is a major function of operator proficiency, judgment, and experience. A good homing DF operator evaluates signal conditions, siting effects, and environmental factors and then defines the effects on homing accuracy.

7.3 NAVIGATION

In many parts of the world, precision, long-range navigational aids such as LORAN, SATNAV, GPS, and TACAN are readily available. However, in other parts of the world, navigational aids are sparse or nonexistent, and radio direction finding is a major navigational aid. An example is Australia and vicinity where LORAN is not available, and SATNAV and GPS coverage is intermittent. Hence, Australia and vicinity is heavily populated by marine and aeronautical MF and HF radiobeacons for navigation by RDF.

Even in the areas covered by modern navigational aids, many locations provide poor performance [2]. For example, the southern part of Florida and the Caribbean experience reduced LORAN-C accuracy due to geometric dilution of precision (GDOP) created by shallow crossing angles between the signal propagation paths. Also, the Bermuda area and the island of Hawaii experience poor LORAN-C performance due to low signal strength. If local thunderstorms occur, LORAN-C performance in Bermuda may become unsatisfactory. The result is that RDF is a major navigational aid in Bermuda, where excellent marine radiobeacon coverage is maintained [2].

A battery-powered RDF may be the only navigational aid available after the primary power fails on an aircraft or vessel. Further, on vessels, lightning strikes or dismasting can knock out LORAN, SATNAV, and GPS antennas and equipment. Clearly, RDF will remain a major navigational aid for some time.

RDF navigational use may be as simple as the acquisition of a single check bearing to verify an estimated location on a line of position (LOP) or initiate a change in course from an established LOP or flight path. More complex RDF navigation may involve resection (back) triangulation, which is discussed in Section 7.4.

The most common and reliable MF RDF navigation emitters are the marine and aircraft radiobeacons. In the VHF band, automatic direction finding systems may be able to navigate using the NOAA weather-radio transmissions if the ADFs are within reliable operating range.

7.4 RESECTION (BACK) TRIANGULATION

The geometry for resection triangulation is shown in Figure 7.2(a). The DF system is located at an unknown position (x_i, y_i) on a horizontal reference plane. Emitters 1 and 2 are located at well-charted positions (x_1, y_1) and (x_2, y_2), respectively. The DF system measures azimuth angles ϕ_1 and ϕ_2, respectively, on emitters 1 and 2, with respect to a reliable established reference such as magnetic north, true north, or grid north. At the DF site, the back bearings ϕ_1' and ϕ_2' are calculated as $|180° - \phi_1|$ and $|180° - \phi_2|$, respectively, and are plotted or computed as LOBs from

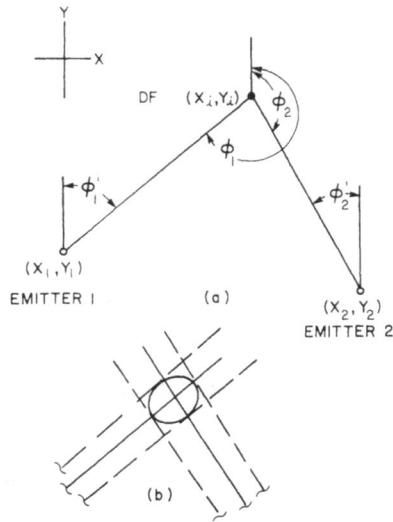

Figure 7.2 Resection (back) triangulation: (a) geometry; (b) error ellipse.

the known emitter locations using the same azimuth reference as that used at the DF. Under error-free conditions, the LOBs along ϕ_1' and ϕ_2' should intersect at the DF site and indicate the DF coordinates (x_i, y_i). Another resection triangulation technique is based on the acquisition of sequential bearings from only one well-charted emitter as the DF system moves along an LOP or flight path. Some DF systems use an automated technique where the emitters and their coordinates are stored in memory as way-points, and the DF position is computed automatically as the bearing angles are measured.

Systematic (bias) bearing errors create an intersection that does not occur at the correct emitter location. In many cases, systematic errors are known, and the back LOBs may be corrected. Random errors "smear" the LOBs as shown in Figure 7.2(b). The dashed lines, which represent bearing uncertainties introduced by random errors, form an area of uncertainty depicted by the ellipse.

DF position uncertainty may be reduced by using more than two emitters for back LOBs. Three back LOBs may form an error triangle of position. As expected, considerable effort has been devoted to analyzing position uncertainties, developing algorithms that estimate the most likely position within an error contour, and evaluating the accuracy of the estimation techniques. Most of the effort has been devoted to direct horizontal triangulation for emitter location estimation, as discussed in Section 7.5. Poirot and McWilliams [3] deal specifically with resection triangulation algorithms for estimating DF position from two bearings and DF position and heading from three or more bearings.

RDF resection triangulation is widely used for marine navigation, especially in those areas of the world where advanced navigational aids are of limited usefulness or nonexistent.

7.5 HORIZONTAL (AZIMUTH) TRIANGULATION

7.5.1 Analytical Techniques

The most basic form of horizontal (azimuth) triangulation is shown in Figure 7.3(a). Bearings from DF sites 1 and 2 are plotted to obtain a position estimate or fix on an emitter E at the intersection of the LOBs. The bearings may be acquired simultaneously from separate DF sites or sequentially from a single DF system on a moving platform. The coordinates of the DF locations are either known *a priori* or measured simultaneously with bearing acquisition.

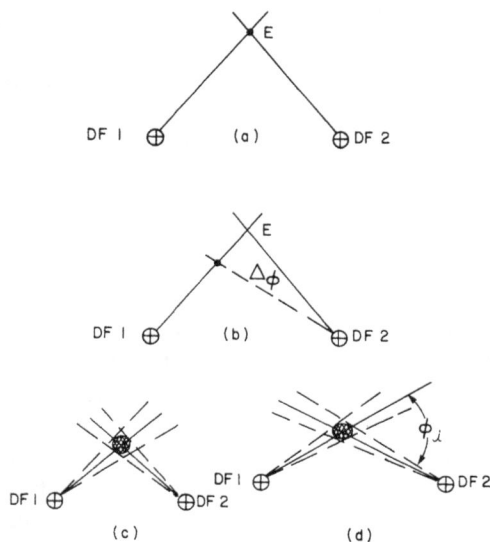

Figure 7.3 Depiction of horizontal triangulation from two DF sites with (a) no error; (b) systematic error; (c) random error; (d) random error.

In Figure 7.3(b), DF 2 contains a systematic error of $\Delta\phi$ on the line of bearing, and, hence, an error in the estimate of emitter position E results. The presence of systematic errors for two-site position locations is difficult to detect, and those errors are a primary reason for using more than two LOBs for position location.

Figure 7.3(c) depicts the situation when random LOB errors exist. The dashed bearing lines indicate the standard deviation or rms error of the azimuth angle. In Figure 7.3(c), the LOBs intersect at a 90° angle providing a minimum circular area of uncertainty with radius r. Figure 7.3(d) shows the general condition where the intersection angle ϕ_i is less than 90°. The area of uncertainty becomes elliptical with a major-axis radius, r', equal to $r/\sin\phi_i$. Along a direct line between the DF sites, r' is infinite, and no useful position information results. This is another reason for increasing the number and dispersion of the DF sites and, hence, the LOBs.

Figure 7.4 illustrates various conditions for the classical azimuth triangulation situation in which the LOBs from three sites at known locations are used to locate the position of the emitter. Figure 7.4 may also apply to a moving DF system acquiring LOBs at three known locations.

In Figure 7.4(a), the desired common-point intersection of three bearings rarely occurs in practice because of error sources. Indeed, a perfect common-point fix does not guarantee that the intersection is the emitter location—as is commonly believed. Two or more of the bearings may be in error, and a common intersection can still occur.

Figure 7.4(b) illustrates the most common case in which the LOBs intersect to form a triangle called the error triangle. Geometry is often used to determine the most probable emitter location. Three of the most used geometric methods are shown in Figure 7.5 where P is the estimated emitter position. The Steiner point method is considered the most accurate of the three. If a four-site LOB position plot is performed, the area of uncertainty is usually an error quadrilateral. The most likely emitter position in the quadrilateral can be estimated in three ways as follows.

1. At the intersection of the two diagonals.
2. At the intersection of the diagonals of the quadrilateral formed by joining the four Steiner points of the four triangles found in the error quadrilateral.
3. At the intersection of the two diagonals of the triangle formed by joining the three Steiner points of three triangles found in the error quadrilateral.

These relatively simple geometric methods are used for DF applications such as marine RDF and ADF back triangulation, ELT and EPIRB beacon location, search and rescue, wildlife tracking, and the location of interference sources and unauthorized transmitters. The position fix may be plotted on a chart or map or computed and displayed on a handheld calculator or field-expedient computer.

In many practical situations, a critical evaluation of the features and typography in the area may reveal the most likely emitter position. For example, tactical VHF communication emitters are usually located on high, open terrain. Operation near power lines and major reradiators is generally avoided. Common sense should be used when defining most likely emitter location from calculations and geometric methods.

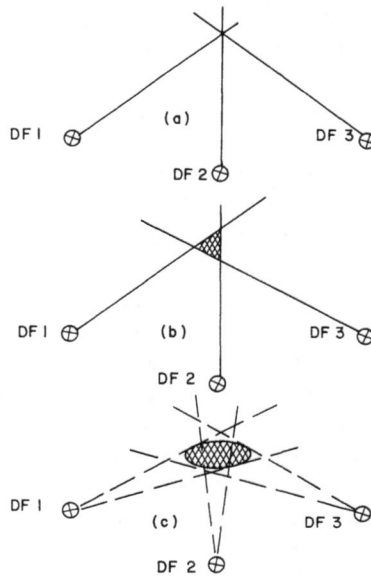

Figure 7.4 Depiction of horizontal triangulation from three DF sites with (a) no error; (b) systematic error; (c) random error.

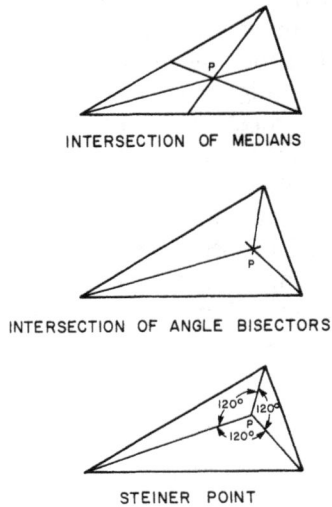

INTERSECTION OF MEDIANS

INTERSECTION OF ANGLE BISECTORS

STEINER POINT

Figure 7.5 Techniques for estimating the most probable emitter position in the error triangle.

Geometric estimation of position based on the error triangle contains deficiencies. For example, the following parameters are not inherent in the estimation: DF site configuration, the distances involved, the standard deviation of the bearings, and the number and quality of the bearings.

The limitations of nonstatistical position estimation methods resulted in numerous investigations to develop statistical estimation techniques. The pioneering work by Stansfield [4] on DF statistical position estimation is especially important. Considerable emphasis has been placed on how to include random errors in the estimation techniques.

In Figure 7.4(c), the random spread of the bearings creates an elliptical area of position uncertainty assuming a Gaussian distribution of the bearing errors. The random bearing error is defined by the standard deviation or rms error. (Many DF system specifications use rms and standard deviation interchangeably.) The size, position, and elliptical probability area are determined by factors such as the number of DF bearings, the range of bearings, and the standard deviation of the errors. For simplification, the elliptical position estimation is usually cited in terms of the radius of an equivalent error circle for a specific probability level that the emitter position will lie in the circle. This descriptor is called circle error probability or CEP.

Statistical position estimation efforts have concentrated on developing algorithms and mathematical models that define the most probable emitter position and the probability distribution of position estimations from the most probable position. The essential elements of the Stansfield method are shown in Figure 7.6. Figure 7.6(a) depicts a set of LOBs from the k DF locations on an emitter at coordinates (x_e, y_e). The (x_i, y_i) coordinates of each DF location are known or measured, and the individual lines of bearings, ϕ_j, are measured. The standard deviation of each individual LOB is known. Figure 7.6(b) shows a detailed view of parameters pertinent to the jth line of bearing. The measured bearing is $\tilde{\phi}_j$; however, it is in error by $\Delta\phi_j$. The parameter l_j is the perpendicular distance between LOB $\tilde{\phi}_j$ and the actual emitter location. Stansfield formulated an algorithm for estimating (x_e, y_e). The algorithm is called the minimum perpendicular distance error method [5]. First, a bearing error model based on using l_j is defined as the sum of the l_j^2 values. This summation is quadratic in the variable (x_e, y_e). Next, the partial derivatives of the summation are derived and set equal to zero. This provides a matrix expression in x_e and y_e as follows [5]:

$$\begin{bmatrix} x_e \\ y_e \end{bmatrix} = \left\{ \sum_{j=1}^{k} \begin{bmatrix} \cos^2\phi_j & -\sin\phi_j\cos\phi_j \\ -\sin\phi_j\cos\phi_j & \sin^2\phi_j \end{bmatrix} \right\}^{-1}$$
$$\cdot \sum_{j=1}^{k} \begin{bmatrix} x_j\cos^2\phi_j & -y_j\sin\phi_j\cos\phi_j \\ -x_j\sin\phi_j\cos\phi_j & y_j\sin^2\phi_j \end{bmatrix}$$

(7.1)

where

ϕ_j = the bearing observed from the jth DF location with respect to a reference such as north;

x_j = the x coordinate of the jth DF location;

y_j = the y coordinate of the jth DF location.

The basic Stansfield algorithm given by Eq. (7.1) was generalized by Smith *et al.* [6] to provide for measurements taken with different DOA accuracies and at different ranges. Each matrix element in Eq. (7.1) is divided by the factor $\sigma_j^2 D_j^2$ where σ_j is the standard angular deviation of the jth bearing and D_j is an estimated distance to the emitter from the jth DF site. The use of an estimated distance does not significantly degrade the accuracy of the position estimate [6].

Ellipse error parameters are calculated based on the following geometric parameter set derived from Reference [6]:

$$\lambda = \sum_{1}^{k} \left(\frac{\sin^2 \tilde{\phi}_j}{\sigma_j^2 D_j^2} \right) \tag{7.2}$$

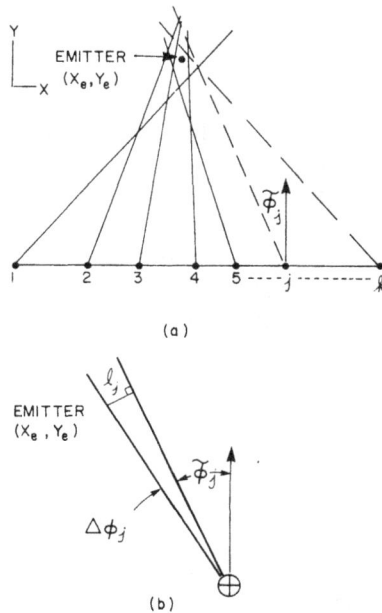

Figure 7.6 Diagram pertinent to the Stansfield algorithm.

$$\mu = \sum_1^k \left(\frac{\cos^2\tilde{\phi}_j}{\sigma_j^2 D_j^2} \right) \tag{7.3}$$

$$\nu = \sum_1^k \left(\frac{\sin\tilde{\phi}_j \cos\tilde{\phi}_j}{\sigma_j^2 D_j^2} \right) \tag{7.4}$$

where

$\tilde{\phi}_j$ = the measured bearing angle from the jth location;
σ_j = the standard angular deviation of the jth bearing;
D_j = the estimated distance to the emitter from the jth DF site;
k = the number of DF locations.

Using the parameters λ, μ, and ν, the error ellipse semimajor, a, and semi-minor, b, axes may be computed from the following expression:

$$a^2, b^2 = 2 \left[\lambda + \mu \pm \sqrt{(\lambda - \mu)^2 + 4\nu^2} \right] \tag{7.5}$$

An elliptical contour of equal probability is defined by

$$\frac{x^2}{a^2} + \frac{y^2}{b^2} = -2 \ln(1 - P) \tag{7.6}$$

where P is the fractional probability that the emitter will lie within the area bounded by the ellipse contour.

The major axis of the ellipse lies along the vector sum of the LOBs through the estimated emitter position. The LOBs are weighted relative to the distance D_j. Figure 7.7 shows representative probability ellipses for a three-station DF net sited in the form of a right-angled isosceles triangle. Bearing standard deviation is the same for all sites, and the error contours are the same probability. The difference in ellipse shapes is due to GDOP, which increases position estimation errors in certain regions because of "poor" LOB angles. For example, when two LOBs cross at a small acute angle (a shallow angle), position estimation accuracy is reduced along the axis of the acute angle, and the probability ellipse is elongated. Wiley [7] presents probability ellipses for five representative DF configurations. In Appendix B of Reference [7], Wiley derives the probability ellipse parameters for a two-site DF net.

Numerous variants of Stansfield's basic algorithm have been formulated. Torrieri [8] applies robust statistical techniques to the Stansfield algorithm and includes the statistical effects of environmental and thermal noise on bearing standard deviation. Torrieri [8] also presents a discussion of the statistical theory of

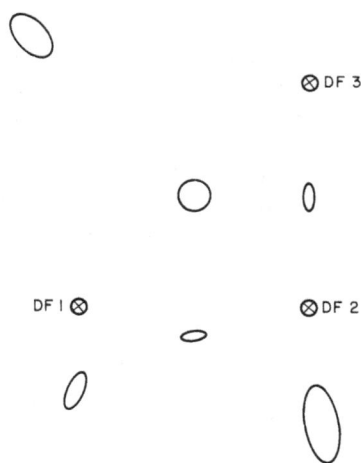

Figure 7.7 Representative probability error ellipses for a three-station DF net.

passive location systems using maximum likelihood or least-squares estimation theory. Blachman [9] derives a simpler version of the Stansfield algorithm by using a complex number representation of the bearing error probability density function. Blachman's approach avoids the need for any special knowledge of a bivariate normal distribution and obviates the requirement for matrix inversion in the computations. Butterly [10] develops position estimation techniques based on a Bayesian approach, which permits available *a priori* information to be included in the estimation process. Foy [11] develops position estimation techniques using a least-sum-squared error solution based on a Taylor series estimation. Springarn [12] reports on the use of extended and iterated extended Kalman filtering to obtain the optimal-filtered emitter estimate. Springarn [12] compares the Kalman filter techniques to the nonlinear least-squares or Gauss-Newton iterative method, which is also discussed. The nonlinear least-squares technique defines position estimation in terms of $\Delta\phi_j$ in Figure 7.6 rather than in terms of l_j. The position estimation function is quadratic in $\Delta\phi_j$ and, hence, nonlinear in x_e and y_e. Sparagna *et al.* [5] present an algorithm for x_e and y_e estimation based on a minimum angular bearing error method. This method uses reiteration and all of the past estimates for "present" x_e and y_e computations. Foy [11] provides a Taylor series approximation to the Gauss-Newton method. Sparagna *et al.* [5] point out that linear approximations have more computation requirements and parameter stability demands than the minimum perpendicular distance estimation technique.

Many triangulation position estimation algorithms typically assume that systematic (bias) errors are not present and only random errors exist. Random errors are usually minimized by averaging over a large number of position estimates obtained from independent bearing measurements. Systematic (bias) errors are more difficult to negate by averaging processes. Estimation techniques that deal with systematic errors have been generated. Poirot and McWilliams [13] introduced a linear statistics algorithm to reduce certain types of systematic errors and to determine the "best fit" mathematical models by examining the raw data for possible systematic error. A limitation of the Poirot and McWilliams algorithm requires a relatively large amount of bearing and DF position data for efficient operation. The algorithm is well suited for an airborne DF system taking numerous bearings along a "good" flight path, e.g., a flight path twice the range to the emitter.

Poirot and Arbid [14] suggest an algorithm based on circulation rather than triangulation to eliminate systematic error. Emitter position is estimated based on intersecting circles instead of intersecting lines. Two major limitations are (1) the requirements for a regimented DF data acquisition path and (2) the sensitivity of the circulation algorithm accuracy to random errors. Poirot and Arbid advise that "tests" for systematic errors be performed following techniques discussed by Poirot and Smith [15].

Mahapatra [16] suggests a method of position location based on the DF system following a path such that the apparent bearing is held constant. The ideal path is a logarithmic spiral. The actual bearing can be acquired by measuring the parameters of the log spiral path. In principle, the systematic error is then eliminated. The obvious disadvantage is the need for the DF system to follow a structured path. Again, airborne DF systems are desirable.

Mangel [17] reports on an operationally simple method for triangulation involving three bearing measurements and two turns of the DF platform along a predetermined path. This technique is a "linearized" version of the logarithmic spiral technique [1]. Systematic errors need not be known, and the effects of random errors are incorporated into the position estimation algorithm. This method is best suited for airborne operations.

7.5.2 Error Descriptors

Passive geolocation errors are described by location probability uncertainties using geometric characterization. For three-dimensional error conditions, the primary error descriptor is an ellipsoidal surface enclosing a minimum volume of a specified error probability. The ellipsoid error is represented by a spherical error probability (SEP) in which the enclosing surface is a sphere rather than an ellipsoid. For two-dimensional error conditions, the generic statistical error descriptor is a two-dimensional ellipse; however, the most useful descriptor is the circular error probability or circular probable error (CPE), which represents a two-dimensional bearing error distribution with the radius of the circle usually encompassing 50% of all the errors.

The circle may represent other probability values because the general notion of SEP and CEP does permit a definition of other spheres and circles with different probabilities.

Range and cross-range error descriptors are often used instead of CEP descriptors [5], especially for airborne stand-off DF applications. A representative airborne stand-off DF scenario is shown in Figure 7.8. Normal conditions produce a low depression angle to the emitter from the aircraft; hence, a two-dimensional position location situation results. The DF system traverses a stand-off path along a cross-range distance CR. The distance R is the perpendicular distance to the emitter from the nominal cross-range path. The ideal CR path is a straight line; however, factors such as political boundaries and the forward line of troops (FLOT) may require a nonlinear path.

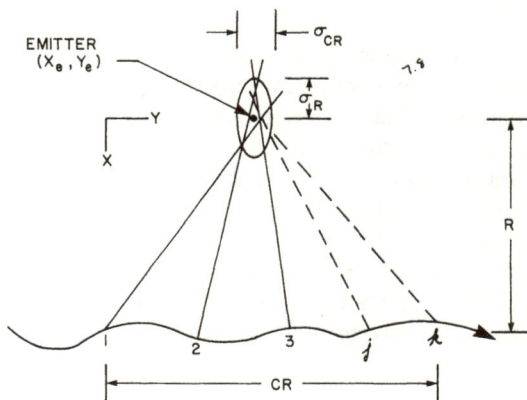

Figure 7.8 Range and cross-range error parameters.

Along the path CR, the DF system acquires k bearings on the subject emitter. The standard deviation or rms bearing error is known. Position estimates (x,y) and CEP parameters are computed using algorithms such as those by Stansfield (minimum perpendicular error) and Gauss-Newton (minimum angular error). The standard deviations in range (σ_R) and cross-range (σ_{CR}) are then derived from the position estimates and CEP parameters [5]. Both σ_R and σ_{CR} are related to CEP. Johnson et al. [18] provide data for computing σ_R and σ_{CR} from CEP data. Figure 7.9, from Reference [18], plots 50% probability CEP, normalized by σ_x, versus the ratio σ_y/σ_x, where σ_x and σ_y are equivalent to σ_R and σ_{CR}, respectively. Gaussian error distribution is assumed. Using Figure 7.9, if $\sigma_R = \sigma_{CR}$, the 50% CEP radius is $1.177\sigma_R$. If $\sigma_R \gg \sigma_{CR}$, the 50% CEP radius is approximately $0.67\sigma_R$.

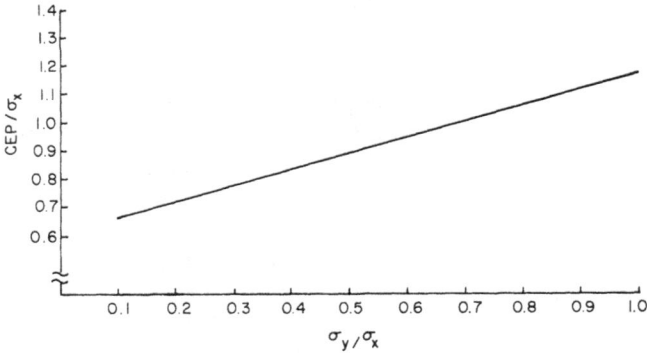

Figure 7.9 Normalized CEP *versus* ratio of the error ellipse standard deviations.

Figure 7.10, from Reference [5], plots normalized standard deviations of range and cross range as a function of the cross-range–to–range (CR–R) ratio for both the minimum perpendicular distance and minimum angular error estimation techniques. The normalization factors for the standard deviations are the rms bearing error and the square root of the number of bearings acquired along the path CR. Therefore, the ordinate value of Figure 7.10 is (σ_R or σ_{CR}/range)(number of bearings along CR)$^{1/2}$/(rms error).

Figure 7.10 indicates that the two error-estimation techniques provide the same error values for CR–R ratios below about 3. For CR–R ratios above about 3, the minimum angular error estimation technique provides the best accuracy. As

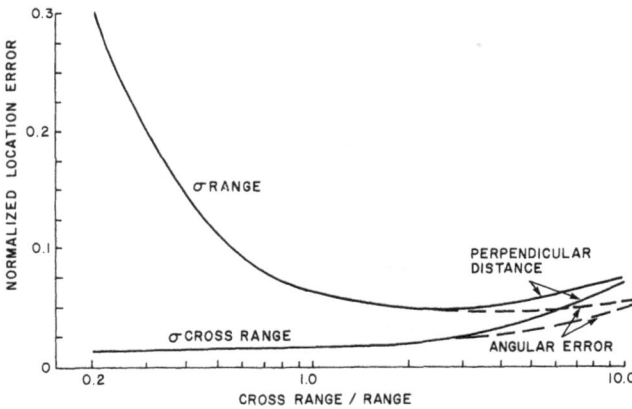

Figure 7.10 Normalized location error *versus* the CR–R ratio (from Sparagna *et al.* [5]).

Sparagna *et al.* [5] state, the minimum perpendicular error-estimation algorithm contains nonoptimal range weighting that reduces accuracy at CR–R ratios above about 3. A CR–R = 3 value produces a minimum σ_R error whereas a CR–R ratio below unity provides relatively large σ_R errors. Therefore, the optimum CR–R ratio range is from about 1 to 3.

Sparagna *et al.* [5] show that the position estimation techniques can produce bias errors generated by the estimation process *per se* even though no bias errors exist in the original measured bearings. Estimation-induced bias errors produce a shift in the CEP center and are most significant below a unity CR–R ratio. For example, the Stansfield algorithm gives a biased estimate of range to the emitter. The exact expression is complicated, but is approximated by

$$(R - R^*)/R = -0.12(\sigma/\phi_s)^2 \tag{7.7}$$

where

$$(R - R^*) = \text{the range bias error;}$$
$$R = \text{the actual range;}$$
$$\sigma = \text{the bearing standard deviation;}$$
$$\phi_s = \text{the angle “swept-out”; the angular extent approximated by the CR–R ratio in radians.}$$

Range bias is generally insignificant. With $\sigma = 5°$ and $\phi_s = 15°$, the bias in range is only about 1%. If the bias becomes significant, Eq. (7.7) may be used to remove the bias.

7.5.3 Direction-Finding Siting Considerations

DF systems siting and net configuration play major roles in increasing triangulation accuracy. For fixed-base siting, multiple DF systems should be uniformly dispersed around the area containing the subject emitters. However, operational and logistic factors often prevent this optimal siting configuration. An alternative siting strategy is to locate the DF sites as far apart as possible, but in close proximity to the area of the subject emitters. Further, the DF sites should be arrayed such that all baselines between sites are nonparallel.

The CR–R error analysis shows that the total DF baseline distance should be approximately three times the estimated maximum distance of the subject emitter field. If close-approach siting is denied and stand-off siting is necessary, the baseline distances between the DF sites should be approximately the same as the distance between the subject emitter area centroid and the centroid of the DF sites, as illustrated by Figure 7.11 [19]. The deployment shown in Figure 7.12 is often used for

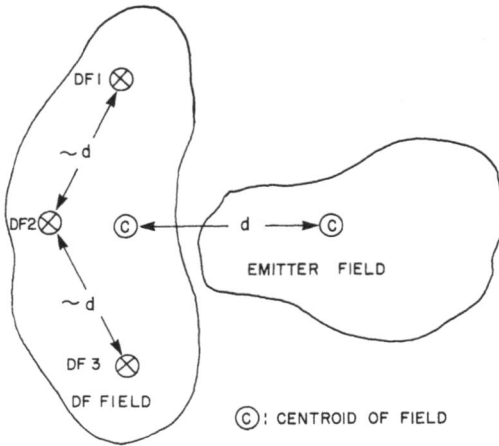

Figure 7.11 DF site deployment based on the centroid method (after Kennedy and Wharton [19]).

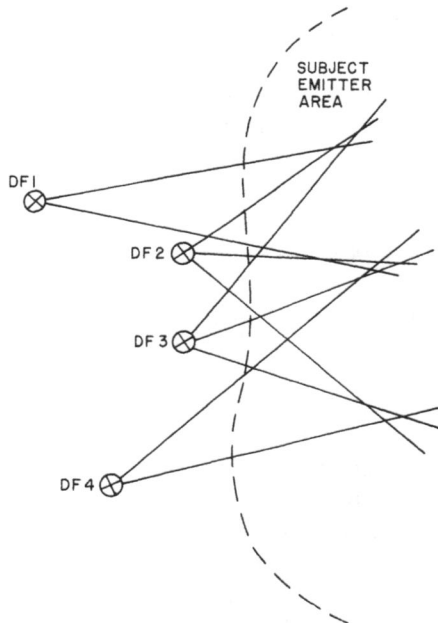

Figure 7.12 DF site deployment for coverage over a broad front.

DF operation over a broad front of emitters. For stand-off airborne DF operations, the CR–R ratio (Figure 7.10) should be maintained at about three, the flight path should be as straight as possible, and the number of independent bearing measurements should be maximized. The angular standard deviation is inversely proportional to the square root of the number of bearings obtained.

7.6 VERTICAL TRIANGULATION

7.6.1 Single Station Location

Vertical triangulation is performed by measuring the azimuth and elevation angles of arrival of an HF sky wave signal [19–23]. The geolocation of the subject emitter may be estimated by knowing, or measuring, the height of the ionospheric refracting layer. Emitter distance may be estimated knowing the refracting layer height and measuring the elevation AOA. Ionospheric conditions are usually determined by an ionosounde integrated into the SSL operation. SSL performance is limited by the accuracy of ionospheric propagation and refracting layer data and their timeliness. Performance is considerably degraded if the sky wave experiences more than one refraction in the ionosphere.

The dominant ionospheric effects are layer tilts and traveling wave disturbances. Because these effects are time variables, real-time or near-real-time ionosounde data are necessary. Ionosounde data are usually measured over the 2- to 20-MHz range by vertical incidence and oblique incidence sounders or, to a lesser extent, by remote ground backscatter methods. Active emitters at known locations in or near the subject emitter field may also be used to obtain ionospheric characteristics. Traveling wave disturbances may be identified by time-series tilt measurements, doppler shift, and by a combination of DOA measurements and doppler. The degree and extent of ionospheric measurements depend on the application and the DF scenario. For example, for emitter distances of less than 100 km, real-time measurements of both midpath ionospheric tilt and traveling ionospheric disturbances are necessary.

Heaps [24] treats the effects of ionospheric variability and irregularity on the accuracy of HF position location with emphasis on SSL techniques. Heaps concludes that the ionospheric effects most likely to degrade SSL accuracy are TIDs and ionospheric tilts and that these factors exhibit a spatial coherence of about 50 km and a temporal coherence of approximately 5 minutes. Therefore, Heaps recommends that multiple ionosoundes be used with SSL systems.

SSL distance estimation error generally exceeds azimuth bearing errors. Distance error involves two independent error sources, elevation angle-of-arrival error and ionospheric ray-tracing error. Elevation AOA error is the lesser error, especially at ranges less than about 300 km. Ray-tracing errors are a function of the accuracy of the ionospheric data, the amount of uncompensated ionospheric tilt, and the robustness of the ray-tracing algorithm.

SSL systems use phase interferometer or pseudodoppler DF techniques. The antenna elements are typically vertical crossed loops that provide good near-zenith coverage for short-range (<100 km) operation.

High-accuracy SSL systems, which have a location accuracy within 5% of the actual distance, use multiwavelength apertures [25–27] and, hence, support wave-front analysis to confine DF measurements to times when the incident wavefront is relatively simple [28]. Small-aperture SSL systems with baselines of one-half wavelength or less are performance limited in multicomponent fields and provide only coarse position location information. Therefore, small-aperture DF systems are not favored for SSL applications unless operational demands, such as mobility, dictate their use.

7.6.2 Airborne AZ/EL Location

Airborne vertical triangulation, called AZ/EL DF, is accomplished by measuring both the azimuth and elevation (depression) angle to a surface-based emitter. Then, knowing the aircraft altitude, the surface-based emitter location is estimated by using geometry to compute an intersection with the surface. Accuracy is a function of the accuracy of the estimated emitter altitude on that of the earth and the magnitude of the depression angle to the emitter. Large depression angles are desired; therefore, AZ/EL DF is a short-range DF technique. Baron *et al.* [29] data indicate that ranges less than 20 miles are required for accurate DF under typical conditions, such as 1° elevation error and 100-ft emitter altitude error for an aircraft at an altitude of 1 mile. For these conditions, the range error at 20 miles is 7 miles.

Dual-baseline interferometers are typically used to measure the two DOA angles simultaneously. AZ/EL DF is generally used at microwave frequencies because of aircraft siting problems at frequencies below microwaves.

7.7 DIRECTION-FINDING NETWORKS

Rapid, efficient emitter geolocation by triangulation requires a well-coordinated DF network. The dispersed DF systems in the net must be interconnected by communication and data links that provide real-time command, control, and data interchange. A net control station (or site) is established to maintain net supervision and coordination and to prioritize DF alerting and tasking. The net control station also performs central data collection and processing functions and acts as the communication and data node to other DF nets or command centers.

For small-aperture DF systems, the dominant net application involves horizontal triangulation and emitter geolocation in a tactical situation. Small-aperture

DF systems are rarely used in large strategic, fixed-site radio surveillance and radio monitoring networks. Therefore, the small-aperture DF network must usually accommodate field-expedient, tactical situations. The level of operation may vary from a totally manual mode to a fully automatic mode. Both mobile and fixed nets exist. An example of a manual, mobile net is a search and rescue effort where bearing data are transmitted by voice over a nonsecure communication link and LOBs are manually plotted on a chart or map. Manual, mobile nets are also common in wildlife tracking applications. An example of a fully automated stationary net is a DF net associated with communication intelligence operations. These DF nets involve numerous, well-configured, well-sited, computer-controlled remote DF stations netted by secure communication and data links.

Tactical, military DF applications require advanced DF networks that include many of the following capabilities.

- Integrated surveillance, intercept, and DF functions;
- Rapid response time, including alerting, tasking, data acquisition, and reporting;
- Mobile, transportable, and stationary capability;
- Automatic, semiautomatic, and manual operation selectable based on operational demands;
- Maximum use of computer control and real-time processing;
- Adaptable, flexible network configuration, especially transfer of net control between DF sites as operational conditions warrant;
- Bidirectional, secure communication and data links interconnecting the DF systems;
- Automated processing for emitter sorting, identification, and geolocation.

REFERENCES

1. Geiser, D.T., "A Method of DF Error Correction," *IEEE Trans. Electromagnetic Compatibility,* Vol. EMC-6, No. 3, October 1964, pp. 38–41.
2. Queeny, T.E., "Direction Finders—The Old and the New," *Navigator,* Vol. 1, No. 1, May–June 1986, pp. 22–26.
3. Poirot, J.L., and G.V. McWilliams, "Navigation by Back Triangulation," *IEEE Trans. Aerospace and Electronic Systems,* Vol. AES-12, No. 2, March 1976, pp. 270–274.
4. Stansfield, R.G., "Statistical Theory of DF Fixing," *J. IEE (London),* Vol. 94, Part IIIA, No. 15, March 1947, pp. 762–770.
5. Sparanga, J., G. Oeh, A. Huber, and G. Bullock, "Passive ECM: Emitter Location Techniques," *Microwave J.,* May 1971, pp. 45–50, 74.
6. Smith, F.W., "Estimation of Emitter Location from Aircraft-Collected Bearing Measurements," SESW-G811, Sylvania Electronics Systems, March 1969.
7. Wiley, R.G., *Electronic Intelligence: The Interception of Radar Signals,* Norwood, MA: Artech House, 1985, pp. 110–112.
8. Torrieri, D.J., "Statistical Theory of Passive Location Systems," *IEEE Trans. Aerospace and Electronic Systems,* Vol. AES-20, No. 2, March 1984, pp. 183–198.

9.   Blachman, N.M., "Position Determination from Radio Bearings," *IEEE Trans. Aerospace and Electronic Systems,* Vol. AES-5, May 1969, pp. 558–560.
10. Butterly, P.J., "Position Finding with Empirical Prior Knowledge," *IEEE Trans. Aerospace and Electronic Systems,* Vol. AES-8, No. 2, March 1972, pp. 142–146.
11. Foy, W.H., "Position-Location Solutions by Taylor-Series Estimation," *IEEE Trans. Aerospace and Electronic Systems,* Vol. AES-12, No. 2, March 1976, pp. 187–193.
12. Springarn, K., "Passive Position Location Estimation Using the Extended Kalman Filter," *IEEE Trans. Aerospace and Electronic Systems,* Vol. AES-23, No. 4, July 1987, pp. 558–567.
13. Poirot, J.L., and G.V. McWilliams, "Application of Linear Statistics Models to Radar Location Techniques," *IEEE Trans. Aerospace and Electronic Systems,* Vol. AES-10, No. 6, November 1974, pp. 830–834.
14. Poirot, J.L., and G. Arbid, "Position Location: Triangulation vs. Circulation," *IEEE Trans. Aerospace and Electronic Systems,* Vol. AES-14, No. 1, January 1978, pp. 48–53.
15. Poirot, J.L., and M.S. Smith, "Moving Emitter Classification," *IEEE Trans. Aerospace and Electronic Systems,* Vol. AES-12, No. 2, March 1976, pp. 255–269.
16. Mahapatra, P.R., "Emitter Location Independent of Systematic Errors in Direction Finders," *IEEE Trans. Aerospace and Electronic Systems,* Vol. AES-16, No. 6, November 1980, pp. 851–855.
17. Mangel, M., "Three Bearing Method for Passive Triangulation in Systems with Unknown Deterministic Bias," *IEEE Trans. Aerospace and Electronic Systems,* Vol. AES-17, No. 6, November 1981, pp. 814–819.
18. Johnson, R.S., *et al.,* "A Computation of Radar SEP and CEP," *IEEE Trans. Aerospace and Electronic Systems,* Vol. AES-5, No. 2, March 1967, pp. 353–354.
19. Kennedy, H.D., and W. Wharton, "Direction Finding Antennas and Systems," Chapter 39 in *Antenna Engineering Handbook,* Johnson, R.C., and H. Jasik, eds., New York: McGraw-Hill Book Co., 1984, pp. 39-1–39-33.
20. Treharne, R.F., "Vertical Triangulation Using Skywaves," *Proc. IREE Australia,* Vol. 28, No. 11, November 1967, pp. 419–423.
21. Treharne, R.F., "Single Station Location: A Brief Survey of the Development of a New System on High Frequency Direction Finding," *Convention Digest IREE Australia,* Melbourne, Australia, 1973.
22. Treharne, R.F., and W.B. Johnson, "Compact Angle Measuring System for HF Skywaves," *Convention Digest IREE Australia,* Melbourne, Australia, May 1971, p. 220.
23. Treharne, R.F., "Non-Military Applications of High Frequency Single Station Location Systems," *J. EEE (Australia),* Vol. 1, No. 1, March 1981, pp. 1–6.
24. Heaps, M.G., "The Effects of Ionospheric Variability on the Accuracy of High Frequency Position Location," Report ASL-TR-0095, White Sands Missile Range, NM: U.S. Army Electronics Research and Development Command, Atmospheric Sciences Laboratory, August 1981.
25. Sherrill, W.M., "A Survey of HF Interferometers for Ionospheric Propagation Research," *Radio Sci.,* Vol. 6, No. 5, May 1971, pp. 549–566.
26. Bailey, A.D., and W.C. McClurg, "A Sum and Difference Interferometer System for HF Radio Direction Finding," *IEEE Trans. ANE,* Vol. ANE-10, No. 1, March 1963, pp. 65–72.
27. Anonymous, "New Transmitter Location System Operational," *Electronic Warfare and Defence Electronics,* August 1978, pp. 61–64.
28. Rice, D.W., "HF Direction Finding by Wavefront Testing," Report 1333, Canada: Department of Communications, Communications Research Centre, March 1980.
29. Baron, A.R., K.P. Davis, and C.P. Hofmann, "Passive Direction Finding and Signal Location," *Microwave J.,* September 1982, pp. 59–76. Erratum in *Microwave J.,* November 1982, p. 141.

Chapter 8
SUBSYSTEM CONSIDERATIONS

The performance of a small-aperture DF system depends on the effectiveness of its subsystem designs; therefore, this chapter has been included to discuss subsystem characteristics and design approaches that significantly impact small-aperture DF performance. Four major subsystems considered are the antenna, the receiver, the DF processor, and displays and controls.

8.1 ANTENNAS

8.1.1 Electrically Small Antennas

Electrically small antennas are inherent in small-aperture DF systems; hence, DF designers and users need to understand the fundamental limitations of electrically small antennas in order to avoid undue performance expectations. Hansen [1], Wheeler [2], Chu [3], and Harrington [4] have defined the fundamental principles and limitations of electrically small antennas. A major principle is that the achievable bandwidth of an electrically small antenna is limited because the minimum Q of the antenna is inversely proportional to the size of the antenna in radian wavelengths. Therefore, the antenna Q increases rapidly, following an inverse cube function, as antenna size decreases, and achievable bandwidth is severely limited. This principle is independent of the method of antenna construction.

8.1.2 Broadband (Wideband) Antennas

Numerous attempts have been made to increase the operating bandwidth of electrically small antennas, including conventional broadbanding techniques such as lumped-constant impedance matching and lumped-constant element loading, for example, inductive loading embedded in the elements and capacitive top-loading for E-field elements. Recently, active coupling and continuous resistive element loading have emerged as viable broadbanding techniques. Active coupling is discussed in Section 8.1.3.

Continuous resistive-loading of an electrically short E-field antenna increases the effective electrical length by altering the current distribution on the element from a triangular to a more sinusoidal distribution. The theory and practice of continuous resistive loading are discussed by Kanda [5]. A fundamental principle is that continuous resistive loading establishes the proper impedance needed to generate an outward traveling wave on the element and, hence, increase the effective electrical length. Continuous resistive loading has been effectively used in small-aperture DF systems [6]. The disadvantages are some loss in sensitivity and increase in element weight. Collateral advantages are a reduction in mutual coupling between arrayed resistively loaded elements and improved linear amplitude and phase response over a broad frequency band, for example, a 25:1 band [6].

Frequency-dependent ferrite inductors, integrated into E-field elements, can provide frequency broadbanding [7]. The permeability, μ, of the ferrite inductors varies inversely with frequency. Since inductance, L, is proportional to μ and inductive reactance is proportional to frequency times inductance, the current distribution tends to remain constant over a broad frequency range. (Rohde and Schwarz, Inc., cites a useful frequency range of 25:1 for their ferrite-loaded HE302 dipole [8].)

Passive, conjugate, impedance-matching techniques for wideband antenna coupling are relatively complex. Fano's limit on broadband conjugate impedance matching [9] shows that the number of components required for the conjugate matching of electrically small antennas to a 50-Ω load is excessive and that the desired phase constant requires stringent component tolerances. Therefore, active antenna couplers are favored for wideband impedance matching.

8.1.3 Active Antennas

An active receiving antenna is defined as a passive antenna element, usually electrically small, that feeds an active device at its output port [6]. The term "active antenna" is also used to describe an antenna with active circuits integrated into the physical structure of the antenna [10–12]; however, for DF purposes, the term "active antenna" refers to antennas with active devices at the antenna output terminals only. In this case, the active circuit acts as a coupling network between the antenna and receiver.

The major objective of an active coupler is to achieve, with an electrically short antenna and the active coupler, a receiving sensitivity equivalent to that of an electrically larger antenna without the coupler. This is accomplished while maintaining wideband low-noise figure and low-distortion operation as discussed in the following sections.

Wideband Operation

A coupler design for wideband operation is a function of the antenna type and the operating frequency parameters of center frequency and bandwidth. Electrically small *E*-field antennas (monopoles and dipoles) may be modeled as a frequency-independent voltage source (Thevenin's source) in series with a large capacitive reactance and a much smaller radiation resistance. An active coupler having a relatively high input impedance and voltage gain provides a bandwidth that is relatively independent of the antenna parameters. A compromise design value for the coupler input impedance is between three and four times that of the antenna element impedance assuming a coupler output impedance level of 50 Ω [6].

Electrically small *H*-field antennas (loops) may be modeled as a frequency-independent current source (Norton's source) in parallel with a large inductive susceptance and a much smaller radiation conductance. An active coupler having a low input impedance and current gain transfers the frequency-independent short-circuit current from the small *H*-field element to the receiver.

The use of active impedance cancellation techniques, such as negative impedance converters (NICs), has been investigated. The concept is to use NICs to cancel the relatively large reactances of the electrically small antennas. Unfortunately, practical implementations exhibit problems such as excessive noise figures [13], instabilities, and crossmodulation and intermodulation products.

Noise Figure

Ideally, the sensitivity of a DF system should be externally noise limited. This reduces the demands on the noise figure of the active coupler. In practice, DF systems are externally noise limited up to about 20 MHz. Above about 20 MHz, the active coupler design must include noise figure reduction techniques. For efficient broadband operation, the active coupler noise figure should approximate the frequency dependence of external noise. Figure 4.7 showed how external noise varies with frequency. An efficient active coupler design has a noise figure *versus* frequency function that varies in the same manner as external noise.

The design and performance of active couplers for electrically small antennas have been investigated [6, 14, 15] and noise figure characteristics have been defined. Theoretical efforts [15] defined the adverse effects of low antenna resistance and high reactance on noise figure. These adverse effects motivate designers to first concentrate on reducing antenna reactance by various passive techniques such as the use of "thick" elements, lumped-constant loading, and continuous resistive loading. When the antenna reactance has been reduced to a manageable level, active couplers may be used effectively.

Distortion

Distortion is minimized by maintaining high linearity and dynamic range in the active coupler. Linearity is a function of the specific design and the effectiveness of the linearization techniques used. Dynamic range is set on the low end by noise and on the high end by crossmodulation and intermodulation. Good design practices for increasing dynamic range are (1) maintain low noise figure, (2) establish a coupler gain that is proportional to frequency, and (3) use linear stages such as push-pull amplifiers and low-noise components [8]. Out-of-band filtering is also advised, especially in high signal density situations. The use of in-band, narrow-band preselection may degrade amplitude and phase tracking integrity. Active couplers require input protection devices to protect against lightning, nearby high power transmitters, and static buildup. Therefore, minimization of the distortion-producing effects of these devices must be included in the coupler design. A good active antenna coupler is designed such that the upper limit of system dynamic range is established by the stages following the coupler.

In spite of numerous design constraints, active DF antennas have been successfully implemented and are being widely employed in operational DF systems. Successful designs typically use a combination of passive and active broadbanding techniques. For example, Anderson and Smith [6] describe an active VHF DF antenna design using a thick dipole element and a balun-fed, common-source FET amplifier functioning as an active coupler. The increased element thickness reduces the element reactance, and the balun provides impedance transformation and unbalanced-to-balanced transmission conversion. (A balanced mode minimizes undesired ground currents in the DF transmission line shield.) The amplitude and phase responses of two antennas described in Reference [6] match to within 0.3 dB and 0.5°, respectively, over the 20- to 120-MHz range. When using two or more active antennas in the same system, the active antennas must be *balanced* over the entire expected frequency and dynamic ranges and the range of environmental variations. Also, the effects of hardware aging on balance should be considered.

An additional advantage of using active couplers in an array is the reduction of currents created by mutual coupling between the array elements because the coupling currents flowing along the elements and transmission lines are attenuated by the nonreciprocal nature of the amplifier in the active coupler. Empirical results indicate that scattering losses in E-field arrays are reduced by the use of active coupling [6].

8.1.4 Direction-Finding Baseline Extension

The sensitivity of DF systems using phase measurements across a baseline is limited if the baseline is electrically short. The total phase difference across the baseline is determined by the physical baseline spacing, d; therefore, an effective electrical extension of d may improve sensitivity especially if performance is limited by the phase-differential measurement. Wiley [16] describes an electrical baseline extension technique consisting of a quadrature hybrid feeding a double-balanced mixer at the output of each antenna. Consider a two-element, phase-differential antenna array, and let the output of antenna 1 be

$$\overline{E}_1 = E_1 \sin\omega t \tag{8.1}$$

and the output of antenna 2 be

$$\overline{E}_2 = E_2 \sin[\omega t + (2\pi d/\lambda)(\sin\phi)] \tag{8.2}$$

where ϕ is the azimuth angle referenced to the perpendicular to the baseline. The phase difference between $\overline{E}_1$ and $\overline{E}_2$ is $(2\pi d/\lambda)(\sin\phi)$. If $\overline{E}_1$ is split into quadrature components

$$\overline{E}_{1i} = (E_1/\sqrt{2}) \sin\omega t \tag{8.3}$$

and

$$\overline{E}_{1q} = (E_1/\sqrt{2}) \cos\omega t \tag{8.4}$$

and recombined in a double-balanced mixer, the resultant signal $\overline{E}_{1r}$ is given by

$$\overline{E}_{1r} = (E_1^2/2)(1/2) \sin2\omega t \tag{8.5}$$

A similar processing technique for $\overline{E}_2$ produces

$$\overline{E}_{2r} = (E_2^2/2)(1/2) \sin[2\omega t + (4\pi d/\lambda) (\sin\phi)] \tag{8.6}$$

The phase difference between $\overline{E}_{1r}$ and $\overline{E}_{2r}$ is $(4\pi d/\lambda)(\sin\phi)$, or the effective electrical length of the baseline has been doubled relative to an unextended baseline. A major disadvantage of the baseline extension technique is that a large SNR is required. The squaring process in the double-balanced mixer creates further reduction in the SNR if the SNR is low. If a low SNR is the basic limitation for an "unextended" baseline, the baseline extension technique can further suppress the SNR, and overall performance is not improved.

8.2 RECEIVER

Small-aperture DF receiver types are a function of the DF technique used and the application. For single-channel, amplitude-response DF systems, the receiver is relatively simple and functions as an RF voltmeter to measure antenna amplitude response. Conversely, for multiple-channel, phase- and time-differential DF systems, the receiver is relatively complex, providing both analog and digital processing and parameter measurements. Over the range from simple to complex DF receivers, the receiver types fall into three major categories as follows:

1. Single-channel receiver
   switched RF
   nonswitched RF;
2. Dual-channel receiver
   switched RF
   nonswitched RF;
3. Triple-channel receiver.

Although the specific DF receiver architectures may vary, certain design characteristics are common for all three receiver categories as discussed in Section 8.2.1.

8.2.1 Common Receiver Design Considerations

Superheterodyne receiver operation is essential for improving sensitivity that is inherently low due to the use of small-aperture antennas. (Nonsuperheterodyne receivers such as crystal video receivers are rarely used for small-aperture DF applications except for conditions in which the signal strength is always relatively large.) At frequencies where the noise figure is internally noise limited, low receiver noise figure operation is desirable.

Multiple-frequency conversion is used in the superheterodyne receiver with high IF frequencies to reduce spurious responses and interference in high signal density environments. An image rejection level of 80 dB is a typical design goal.

Frequency tuning by digitally controlled *frequency synthesizers* prevails in current small-aperture DF receivers. Fast, accurate, stable frequency set-on is obtained. Indirect frequency synthesis is favored over direct frequency synthesis because direct synthesis generates more spurious signals and is relatively expensive to implement if fine frequency resolution is required [17]. Indirect synthesis uses the well-known voltage-controlled oscillator (VCO), phase-locked loop (PLL), and phase detector plus filter configuration. Indirect synthesis derives its stability and accuracy from an internal reference oscillator. Typical accuracy values range from 1×10^{-6} per day for a standard temperature-compensated oscillator to 1×10^{-10} per day for oven-controlled oscillators. These accuracy levels provide adequate frequency set-on and measurement performance under nominal conditions. (The advent of a fully operational global positioning satellite system will afford order-of-magnitude improvements in external time and frequency synchronization capability [18].) Because many DF techniques and algorithms require a measure of operating wavelength to compute DOA, operating frequency must be accurately generated and precisely measured.

A *digitally controlled* superheterodyne DF receiver provides rapid frequency tuning, programmable gain and AGC characteristics, selectable IF and baseband bandwidths, demodulation selection, and amplitude and phase tracking correction. The operational convenience of digital control is very significant. Operation of a manual (analog) DF system is labor intensive, and operator overload is not uncommon especially in high signal density conditions where manual frequency tuning can be difficult. The use of digital frequency selection and preset memory affords a major reduction in operator labor and allows the operator more time to deal with DF evaluation and measurement tasks.

Digitally controlled AGC is especially attractive for DF applications that require receiver gain to be constant when DF data samples are being obtained [19]. Digitally programmed AGC stabilizes the gain for the DF measurement period. Analog AGC circuits change the gain of variable gain RF and IF amplifiers resulting in phase nonlinearities. Digital techniques provide (1) control over RF and IF amplitude and phase tracking and (2) improved manual gain control (MGC) performance.

The receiver *input impedance* should have a nominal resistive value of 50 Ω. A typical voltage standing wave ratio (VSWR) level is 3:1. A stable resistive input impedance is essential to providing constant resistive termination and, hence, to maintaining balance in the prereceiver RF circuitry.

For the *RF-switched* modes of DF operation, the RF switches should provide rapid switching, minimum insertion loss, and maximum isolation and dynamic range while minimizing switching-transient interference in the frequency band of interest. For MF through lower VHF operation, "quiet" broadband RF switching is a particularly difficult design situation. (Randall [20] presents a design approach for "quietly" switching MF through lower VHF signals.) Above about 100 MHz, the effects of switching transients are less severe and relatively simple RF switches may be used. (Harris [21] describes an RF switch used in a pseudodoppler DF system operating at 144 MHz.)

A receiver *beat frequency oscillator* is mandatory for aural-null DF techniques.

8.2.2 Specific Design Considerations

Dynamic range requirements are a function of DF technique and application. For homing DF systems, the RF signal level may vary 150 dB or more over a homing traverse. A dynamic range of 150 dB requires RF attenuator step switching to keep the DF receiver operating in the linear region. A typical dynamic range for a DF receiver without RF-switched attenuation is about 80 dB. For navigational DF systems operating on well-channelized, cooperative transmissions in a stand-off mode, the dynamic range requirements may not be as severe as the homing situation. The dynamic range requirements for tactical DF systems may vary depending on the signal density level and amplitude at the DF site. If the DF receiver uses analog-to-digital (A/D) conversion, the dynamic range may be limited by the A/D converter. The dynamic range for A/D conversion is given by [22]

$$DR = 20 \log2^n = 6n \quad dB \tag{8.7}$$

where DR is the dynamic range and n is the total number of bits in the A/D conversion. For $n = 12$ and 14, the dynamic ranges are 72 and 84 dB, respectively. A representative dynamic range without A/D conversion is 80 dB; therefore, 16-bit conversion may be necessary to avoid dynamic range limitation by the A/D conversion.

For DF systems, the upper limit of dynamic range is usually specified using the third-order intermodulation intercept point rather than a gain compression point. Third-order dynamic range is the decibel difference between the noise equivalent level and the input level of two equal amplitude in-band interfering signals that produce third-order intermodulation products equal to the noise equivalent level, NE, where NE is given by

$$NE = NF + 10 \log kT + 10 \log B \tag{8.8}$$

where
   NF = the system noise figure in dB;
   k = Boltzmann's constant (1.38×10^{-23} joule/Kelvin);
   T = equivalent noise temperature in Kelvins (290 K is room temperature);
   B = equivalent noise bandwidth in Hz.

At $T = 290°K$, kT is 4×10^{-21} watts per Hz bandwidth.

Figure 8.1 shows a specific example of a third-order intercept computation. Noise figure tends to increase as the third-order intercept point increases as illustrated by Figure 8.2, which uses representative performance parameters. Clearly, a design compromise exists between reducing noise figure and decreasing the third-order intercept level. The importance of third-order intercept performance increases as the signal density and level increase.

The requirement for *amplitude and phase matching and tracking* in dual-channel and triple-channel receivers is a strong function of DF technique. For example, phase and time response DF techniques generally require a high degree

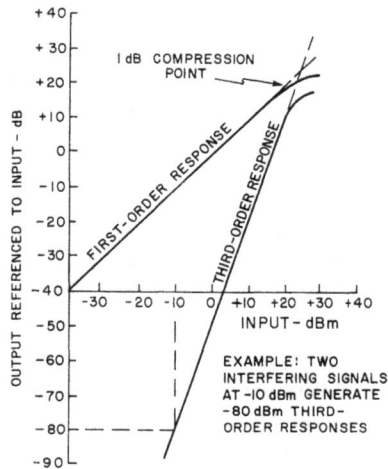

Figure 8.1 Example of a third-order intermodulation computation.

of amplitude and phase matching and tracking. Hence, the receiver local oscillators are derived from a common synthesized frequency source to ensure identical tuning and phasing. If amplitude and phase correction and compensation techniques are used, the requirement for amplitude and phase matching may be alleviated. In this case, the emphasis should be placed on maintaining amplitude and phase stability over time and environmental conditions.

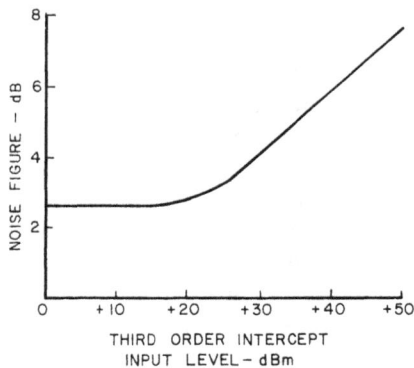

Figure 8.2 A representative plot of noise figure *versus* third-order intercept input level.

8.3 DIRECTION-FINDING PROCESSOR

The basic function of the DF processor is to (1) derive signal DOA and supporting data, such as signal amplitude, from the receiver output, and (2) condition the data for display read-out, storage, and transmission to a remote source. The DF processor may have numerous interfaces such as interfaces with the receiver, displays, data links, supporting peripheral devices, and the controller. In some DF systems, the DF processor and display merge into one unit. Several examples are the triple-channel Adcock–Watson-Watt system of Figure 5.36 and the probed line DF of Figure 5.22.

The complexity of the DF processor is determined by the DF technique and application. In general, DF systems operating on cooperative signal sources in a well-controlled frequency band require the least amount of DF processing. Conversely, DF systems operating on uncooperative, often hostile, signal sources in a dense uncontrolled electromagnetic environment require considerable DF processing. A relatively complex DF processor may perform the following functions:

1. Receiver output A/D conversion;
2. Digital data conditioning;
3. Data transformation and storage;
4. Parameter measurement;
5. Direction-of-arrival computation;
6. Supporting data measurement and computation;
7. Data input-output for peripheral devices;
8. Data conditioning for remote transmission;
9. Controller interface support.

(Figure 6.20 shows a representative DF processor functional block diagram.) Primary DF processor design requirements are computational accuracy, speed, robustness, and efficiency. Accuracy is essential because small-aperture DF error budgets are driven by basic errors and cannot afford additional errors introduced by the DF processor. Parameter measurements of amplitude, phase, and time must be sophisticated, especially for low SNR operation. Computational speed is often a compromise between DOA updated time and interface bus data rate. To reduce computation time, algorithms are truncated; however, the approximations created by the truncation must not compromise the accuracy of the DF data.

A DF processor can have a major impact on system sensitivity. An example is the effect of fast Fourier transform processing on sensitivity as shown by Table 8.1, which was prepared using data presented by Floyd and Taylor [23]. Table 8.1 provides a sensitivity budget for an operational double-baseline, dual-channel interferometer with and without FFT processing. With no FFT processing, the system is assumed to perform as a phase interferometer with direct, nonintegrated, analog phase differential detection. Table 8.1 shows that the sensitivity improvement is 33 dB when FFT processing and integration are used. The 2-s integration time is consistent with typical transmission durations for tactical communication as shown by Table 10.1.

Table 8.1
Representative Sensitivity with and without FFT Processing*

Sensitivity Factor	Sensitivity Increment (dB)		
	without FFT		with FFT
Reference noise level (1.0 Hz BW)	−174	dBm	−174
System noise figure	+7		+7
Equivalent noise bandwidth (10 kHz BW)	+40		+40
Time windowing (weighting)	—		+3.5
FFT processing	—		−27
Differential phase detection	+3		+3
Coherent frequency cell summation	—		−12.5
Noncoherent azimuth cell sum			
(four sets of 0.5-s data)	—		+3
Baseline spacing loss			
(at operating frequency)	+6		+6
Required power level for 0-dB SNR	−118	dBm	−151
Output SNR for 1° rms error		38 dB	
Required power level for 1° rms error	−80	dBm	−113
Output SNR for 3° rms error		29 dB	
Required power level for 3° rms error	−89	dBm	−122

*From Floyd and Taylor [23].

8.4 DISPLAYS AND CONTROLS

Improvements in display technology have had a major impact on small-aperture DF performance. The large, bulky CRTs used by the early DF systems have been replaced by low-power, -weight, and -volume displays using solid-state devices and thin-film electroluminescence. A wide variety of displays has been developed to present supporting data such as signal strength, statistical parameters, and data confidence level and quality. Digital read-out of operating frequency is a common feature. Control and display improvements have employed hand-held terminals and field-expedient minicomputers to enhance operational efficiency and performance significantly.

Analog displays continue as the standard for certain DF techniques and applications. Homing DF systems based on amplitude response typically use either aural-null techniques augmented by a signal strength meter or a left-right meter display. Avionics ADF systems use either a mechanically controlled bearing pointer or an integrated bearing pointer and magnetic indicator (i.e., a radio magnetic indicator). The MF and HF marine RDFs generally use the conventional CRT and "propeller" bearing display. Some marine VHF DF systems use a hybrid combination of digital bearing display and LEDs configured around the circumference of a 360° compass rose.

Instantaneous DF techniques use CRT-based rho (ρ), phi (ϕ) displays of AOA *versus* signal strength. The ρ,ϕ polar display may be augmented by a rectangular AOA *versus* frequency display to better characterize frequency-hopping spread-spectrum transmissions.

Tactical intelligence collection DF systems employ a wide variety of display techniques to present the interrelationships between frequency, azimuth and elevation angles, amplitude, polarization, and time parameters in both analog and digital formats. The display of statistical measures of the signal of interest is becoming common, and techniques for using statistical measures to select DF operating modes in real time are gaining operational status.

The trend in small-aperture DF control technology is toward minimizing the use of knobs, buttons, and switches and maximizing the use of keyboards with tactile membrane switches and displays offering touch-screen menu selection. Internal, local, and remote controls use standard data interface protocols such as RS-232C and IEEE-488. Military systems typically use data interface standard MIL-STD-1553 or a current upgraded version.

The dominant design considerations for small-aperture DF control and display subsystems are operational factors such as weight or volume, form factor, power consumption, human factors performance, and display efficiency. One aspect of display efficiency is a measure of the displayed data per unit panel space assuming other parameters such as weight and volume remain constant. Displays and controls should, whenever possible, be "user-friendly." The DF system designer should bear in mind that the user, such as an aircraft pilot or a boat skipper, has other duties to perform while operating the DF systems.

REFERENCES

1. Hansen, R.C., "Fundamental Limitations in Antennas," *Proc. IEEE*, Vol. 69, No. 2, February 1981, pp. 170–182.
2. Wheeler, H.A., "Small Antennas," *IEEE Trans. Antennas Propag.*, Vol. AP-23, No. 4, July 1975, pp. 462–469.
3. Chu, L.L., "Physical Limitations on Omni-Directional Antennas," *J. Appl. Phys.*, Vol. 19, December 1948, pp. 1163–1175.
4. Harrington, R.F., "Effect of Antenna Size on Gain, Bandwidth, and Efficiency," *J. Res. NBS*, Vol. 64D, January–February 1960, pp. 1–22.
5. Kanda, M., "A Relatively Short Cylindrical Broadband Antenna with Tapered Resistive Loading for Picosecond Pulse Measurements," *IEEE Trans. Antennas Propag.*, Vol. AP-26, May 1978, pp. 439–447.

6. Anderson, H.W., and R.S. Smith, "Active DF Antennas," *RF Des.,* November 1989, pp. 62–68.
7. Czerwinski, W.P., "Wideband Antenna with Frequency Dependent Ferrite Core Inductor," PAT-APPL-6-028 934/WY, Washington, D.C.: Department of the Army, patent filed 11 April 1979, AD-D006277/8.
8. Anonymous, "Antennas for Radiomonitoring and Radio Reconnaissance," *News from Rohde and Schwarz—Special,* 1990, pp. 45–62.
9. Fano, R.M., "Theoretical Limitations on the Broadband Matching of Arbitrary Impedances," *J. Franklin Inst.,* Vol. 249, January–February 1950, pp. 57–83, pp. 139–154.
10. Meinke, H.H., "Active Antennas," *Nachrichtentech Z.,* Vol. 19, 1966, p. 697.
11. Meinke, H.H., "Electrically Small Active Receiving Antennas," *Proc. ECOM-ARO Workshop Electrically Small Antennas,* Ft. Monmouth, New Jersey, March 6–7, 1976.
12. Ramsdale, P.A., and T.S. MacLean, "Active Loop-Dipole Aerials," *Proc. IEE,* Vol. 118, 1971, pp. 1698–1710.
13. Bahr, A.J., "On the Use of Active Coupling Networks with Electrically Small Receiving Antennas," *IEEE Trans. Antennas Propag.,* Vol. AP-25, No. 6, November 1977, pp. 841–845.
14. Rohde, U.L., "Active Antennas," *RF Des.,* May–June 1981, pp. 38–42.
15. Santini, R.A., "Active Antenna Performance Limitation," *IEEE Trans. Antennas Propag.,* Vol. AP-30, No. 6, Nov. 1982, pp. 1265–1267.
16. Wiley, R.G., *Electronic Intelligence: The Interception of Radar Signals,* Norwood, MA: Artech House, 1985, pp. 93–94 and Figure 4-7.
17. Brewerton, D.B., and N. Urbanila, "Synthesizer Primer: Defining the Elements of Good Design," *Microwave RF,* June 1984, pp. 79–86.
18. Daly, K.C., and G. Smith, "The State-of-the-Art in Time and Frequency Measurement," *Microwave J. 1990 State of the Art Reference,* September 1990, pp. 149–154.
19. Drapac, M.J., "Advances in ESM Receiver Technology," *J. Electron. Defense,* April 1987, pp. 83–86.
20. Randall, K.S., "A Lowband RF Quiet Switch," *RF Des.,* July 1990, pp. 29–40.
21. Harris, T., "A Simple Low-Cost RF Switch," *RF Des.,* July 1989, pp. 53–54.
22. Tsui, J.B.-Y., *Digital Microwave Receivers: Theory and Concept,* Norwood, MA: Artech House, 1989, p. 17.
23. Floyd, P., and J. Taylor, "Dual-Channel Space Quadrature-Interferometer System," *Microwave System Design Handbook,* 1987, pp. 133–148.

Chapter 9
CALIBRATION AND TEST OF DIRECTION-
FINDING SYSTEMS

9.1 METHODS

Test and calibration of small-aperture direction-finding systems involves a five-step process as follows:

1. Bench test and calibration;
2. Shielded room and anechoic chamber test and calibration;
3. Calibrated site test and calibration;
4. Installed test and calibration;
5. Operational test and evaluation.

The essential elements of each of these five steps are discussed in the following sections.

9.2 BENCH TESTS

Bench test and calibration are performed in the laboratory with the DF antenna system replaced by a dummy, or simulated, antenna. The major objectives are to test and calibrate the receiver, processor, computer, and control-display parameters associated with the following representative functions:

1. Sensitivity (direct RF input used)
   - IF
   - video-baseband
   - minimum discernible AOA display and read-out
   - AGC threshold
   - bit error rate (FFT processing)
   - squelch break level;

2. Noise figure;
3. Selectivity
   • RF
   • IF;
4. Bandwidth
   • RF
   • IF
   • video;
5. Dynamic range;
6. Image rejection;
7. Co-channel and adjacent channel rejection;
8. Spurious response rejection;
9. Cross modulation and intermodulation;
10. AGC characteristics;
11. Audio output power and distortion;
12. Amplitude and phase response (as a function of input signal frequency, amplitude, and duration);
13. Time-domain pulse response;
14. Scanning parameters (scanned receiver);
15. Settling time;
16. Parameter computation response time;
17. Parameter measurement resolution, accuracy, and stability (as a function of frequency, time and amplitude, and the effects of frequency mistuning);
18. Control and displays interface performance.

Some classes of DF systems have well-defined bench test and calibration procedures. For example, Special Committee 146 of the Radio Technical Commission for Aeronautics (RTCA) published Document No. RTCA/DO-179, dated May 1982, entitled "Minimum Operational Performance Standards for Automatic Direction Finding (ADF) Equipment." Section 2 of this document is devoted to bench testing [1].

The bench tests should include measurement of the desired equality of performance of two or more receiver channels when such performance is called for by the system technique used. Measurements are usually made by using a single RF test source that is split into N sources, where N is the number of channels including the sense channel if one is used. The relative amplitudes or phases of the N sources are then separately controlled to simulate signal angle-of-arrival conditions and DF antenna and baseline effects. For example, a signal simulator for the triple-channel Adcock–Watson-Watt system shown in Figure 5.36 may consist of a single RF source feeding a three-way power splitter. Each of the three splitter outputs is applied to calibrated, variable attenuators and phase shifters to develop simulations of the antenna output voltages shown in Figure 5.36.

9.3 SHIELDED ROOM AND ANECHOIC CHAMBER TEST AND CALIBRATION

Controllable, deterministic electromagnetic fields can be generated in shielded rooms and anechoic chambers. These facilities are used to measure the performance characteristics of small-aperture DF systems including the antenna system. The antenna system should have dimensions that are small relative to the dimensions of the enclosure. At MF and HF, the requirement for small antennas restricts enclosure testing to those systems using low-profile loop antennas. This includes the MF and HF marine and aircraft RDF and ADF systems.

Standard electromagnetic fields may be established in a conductive, shielded enclosure by using a standard signal generator driving a calibrated test transmission line terminated in a resistance equal to the line's characteristic impedance [1–3]. The electromagnetic field, which may be calibrated and controlled, simulates free-space, far-field, vertically polarized conditions. References [1] and [2] describe installation setup procedures and field computation algorithms. The field strength may be computed to an accuracy of about 1% using methods given by Haber [4]. Reference [1] presents an alternative field computation technique. In practice, the shielded room technique is limited in frequency. An upper frequency limit in the lower HF band is typical.

Anechoic chambers are also used to test DF systems in standard fields. The chamber dimensions should be large enough to simulate far-field conditions. Chamber dimensions of 10λ are desirable if the radiating source aperture is small relative to a wavelength. If an electrically large radiator is used, the chamber dimensions should be greater than $2D^2/\lambda$, where D is the largest dimension of the aperture and λ is the operating wavelength. In practice, the useful frequency range for most anechoic chambers begins in the VHF band.

Shielded room and anechoic chamber testing is very effective for measuring antenna parameters such as response pattern characteristics, pick-up factor, and field strength sensitivity effects.

Scale model testing [5] is a reliable, cost-effective method for measuring antenna response patterns, and the anechoic chamber is well suited for testing of small-aperture DF techniques on scale models of the host platform. The frequency range from MF through VHF is compatible with practical scaling ratios. For example, 60:1 is a reasonable scaling ratio for VHF DF antennas on helicopters.

The advent of efficient moment method codes has diminished the need for scale model testing of experimental DF antenna placements on host platforms. Excellent correlation exists between moment method computations and anechoic chamber empirical data. The Appendix discusses moment method codes suitable for DF system analysis.

Suitable shielded rooms and anechoic chambers are excellent test and calibration facilities for small-aperture DF systems, especially if the systems are portable or transportable and can be moved from site to site. Basic system error sources may be readily measured and used to create correction tables.

9.4 CALIBRATED OPEN-FIELD TESTING

Many DF systems go directly from controlled enclosure testing to the host platform for follow-on installation and operational test and calibration. However, many other DF systems that are not compatible with enclosure testing require additional testing before operational use. A calibrated, open-field test facility bridges the gap between production and operational use for these systems. Procedures for calibrated, open-field testing are presented in publications by Harrington [6, 7] from the Watkins-Johnson Co.

A field site acceptable for DF system calibration and testing should possess all of the characteristics of a "good" ground-based DF site. Table 4.4 lists these characteristics. Further, the test area should be at least 15λ long and 10λ wide at the lowest test frequency in order to reduce the field strength of reflected signals [6]. If reduced test site dimensions must be used, major reflecting and reradiating objects should be duplicated on the opposite side of the path of propagation. Figure 9.1 shows a field configuration with limited dimensions and a duplicated reradiator. Figure 9.2 depicts a "large" dimension test site with reflectors and reradiators in a safe region. According to Harrington [6], the influence of site reflectors and reradiators on test data may be identified by rerunning tests with the DF antenna at a new position, for example, at least a full wavelength in any direction. Disruptive reradiators and reflectors are present if the average angle-of-arrival error changes by more than several degrees.

The test RF source (TX) may be either fixed or movable. For a movable RF source, surveyed test points are used at increments along a 360° circle about the DF location. For a fixed point RF source, the DF antenna is rotated to obtain response data. If the RF test source is mobile, each test transmitting position should meet the range criteria depicted in either Figure 9.1 or Figure 9.2, depending on the type of site.

The field strength at the DF antenna is measured for each test transmitting position as a function of frequency and, if needed, polarization. (The DF antenna is, of course, removed for the field strength measurements.) The effective radiated power of the test source must be kept at a constant level. Field strength measurements should be performed using standard procedures [8, 9]. For a fixed frequency, the variation of field strength, relative to the average, as a function of azimuth should not exceed 1 dB after corrections are made for RF power source level variations. Variations greater than 1 dB usually indicate the presence of reradiation or reflection levels that can compromise accurate test and calibration.

After field strength calibrations are completed and the DF site judged acceptable, the subject DF system is installed, and performance data are acquired. Data

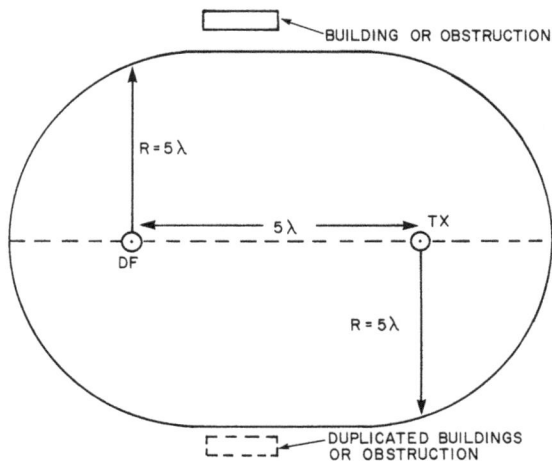

Figure 9.1 Small-field DF test site with obstructions duplicated. (From Harrington [6] Figure 1. Reprinted with permission.)

may include field strength sensitivity, response patterns, signal response time, polarization response, *et cetera*. However, the primary data are bearing *versus* azimuth, which must be statistically analyzed to obtain performance parameters and develop error correction data. The analysis procedure formulated by Harrington [6] is a step-by-step process as summarized below:

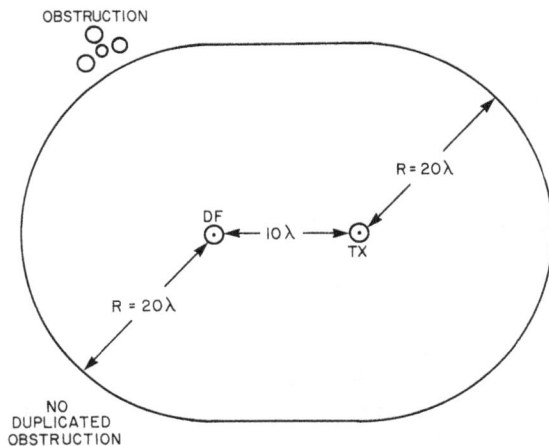

Figure 9.2 Large-field DF test site with obstructions in safe region. (From Harrington [6], Figure 2. Reprinted with permission.)

Step 1

At each test frequency, the measured line of bearing is subtracted from the true LOB to obtain bearing differentials ΔLOB.

Step 2

The standard deviation value of the bearing differentials is computed using compensated ΔLOB values. Reradiators and reflectors may introduce bias or offset in the data. The bias may be computed by averaging all the ΔLOBs at a fixed frequency. Compensated LOBs are derived by subtracting the ΔLOB average from each ΔLOB value. Standard deviation (SD) is calculated from

$$ \text{SD} = \sqrt{\sum_{i=1}^{n} \frac{(x_i - \bar{x})^2}{n - 1}} \tag{9.1} $$

where x_i is the ith ΔLOB, $\bar{x}$ is the average of the ΔLOBs, and n is the total number of LOB data points. (DF practitioners and manufacturers often cite the SD value as the rms accuracy.) The SD for LOBs over 360° in azimuth around the DF antenna is a good measure of DF accuracy at the particular test frequency.

Step 3

Performance of a statistical analysis may be necessary to compare the SD of a data set with a procurement specification or the SD of another data set that may be historical. For comparison purposes, the SD data should be calculated from a large number of individual data points. At VHF and UHF, five frequencies and thirty-six 10° azimuth increments are used to compute an overall system standard deviation. However, the character and size of the data set are based on the needs of the DF system engineer or user. For example, polarization data may be included as an independent variable in the SD computations.

The statistical comparison between the SD values acquired from the test data and the specification or archive data set is performed by using the chi-square (χ^2) test, which may be viewed as a tolerance test of the SD based on a specified confidence level. A typical confidence level for DF test data is 95%, which means that only 5% of the satisfactory, acceptable data will fail the chi-square test. The probability of accepting an unsatisfactory data set is reduced to near 0% when 72 or more data values are used. Using five separate frequencies and 10° azimuth intervals, the number of data points is 180.

The chi-square parameter χ^2 is computed by

$$\chi^2 = (n - 1)(SD_c)^2/(SD_s)^2 = \gamma SD_c^2/SD_s^2 \tag{9.2}$$

where

n = the number of data points;
SD_c = the computed standard deviation;
SD_s = the desired or specified standard deviation.

The parameter n should be larger than 30. The acceptance χ^2 level is χ_o^2, which may be obtained from standard mathematical reference tables that plot $\gamma = n - 1$ *versus* χ_o^2. If the calculated χ^2 value is less than χ_o^2, the DF system performance is judged acceptable.

Step 4

If needed, DF error correction factors can be computed using the raw ΔLOB data. For a given polarization, ΔLOB data are grouped by frequency and azimuth, and then the grouped data are averaged across time to obtain a correction factor for each frequency and each azimuth increment.

Step 5

The azimuthal increments may be too coarse for high integrity error correction; therefore, calculation is necessary of the interpolation algorithms that estimate values between the azimuthal increments. Linear and curved interpolation algorithms can be computed. Curve smoothing may be reiterated a number of times until a desired degree of smoothness or cohesiveness is obtained.

A variety of DF operating parameters may be involved in the calibration process. One important parameter is the vertical response pattern of the DF system even though the horizontal response is usually of primary interest. The vertical response pattern may be measured in conjunction with elevation angle-of-arrival measurements as discussed in the next paragraph.

Elevation angle-of-arrival characteristics are measured using aircraft-towed multifrequency transmitters. Barker [10] reports on a system operating from 2 to 100 MHz. This system, called the Xeledop system, is capable of measuring both vertical and horizontal polarization. Aircraft position is determined by ground-based tracking systems. The future use of GPS receivers on the aircraft may negate the need for a separate ground-based tracking system.

Open-field testing for azimuthal DF data should avoid errors introduced by undesired variations in the elevation patterns of either the target transmitter or the DF antenna. Mitigation techniques have been discussed in Section 4.4 of Chapter 4.

9.5 INSTALLED TEST AND CALIBRATION

Installed test and calibration are performed with the DF system installed on the host platform or at the operating site. Installed test and calibration of ground-based systems are often performed using calibrated, open-field sites. First, the DF system is installed and tested on a well-controlled ground plane and then installed on the host platform and tested again. This sequential testing delineates the effects of the host platform and separates instrumental, environmental, and site errors. DF systems that use stored amplitude and phase response data for angle-of-arrival measurement generally obtain the data to be stored from installed testing on calibrated open-field sites.

Many mature DF systems with high volume production and reliable experience factors bypass open-field calibration and go directly to installed test and calibration. Examples are aircraft and marine RDF and ADF systems.

Installed test and calibration are often tailored to the application and host platform. For example, installation test and calibration of a marine VHF ADF system may emphasize the homing function. A VHF target transmitter located in the far-field directly off the bow is used. If the bearings on the source are within $\pm 5°$ of the boat centerline, the DF system is judged satisfactory for homing. At other azimuths around the boat, bearing errors may be as large as $\pm 20°$ and still be acceptable.

On aircraft and vessels, installed test and calibration measure deviation errors as a function of frequency, azimuth angle, and, sometimes, elevation angle. Correction tables are prepared from the deviation errors. Some DF systems incorporate the correction data into electronic compensation processes that automatically correct for deviation.

Installation test and calibration are also used to determine if the DF system is electromagnetically compatible with other electronic systems installed on the host platform.

9.6 OPERATIONAL TEST AND CALIBRATION

Operational test and calibration are (1) performed with the DF system in an operational environment and (2) conducted by the operating personnel assigned to the DF system. The main objective is to evaluate the entire DF system performance when operated by the assigned personnel engaging real world, authentic signals of interest. The signals of interest may be either (1) cooperative, local sources or (2) cooperative or uncooperative remote sources. An aircraft-towed test transmitter, such as the Xeledop system [10] is an example of a cooperative local source. Uncooperative remote sources are typically at a known fixed location, operate on a more or less regular schedule, and are identifiable. Remote sources at fixed locations are excellent "calibration beacons" especially for HF DF on sky wave transmissions. At a fixed site, a database of "check" bearings obtained over a variety of angles of arrival and frequencies over an extended period is invaluable for identifying deteriorating performance and decreasing site integrity.

REFERENCES

1. Special Committee 146, *Minimum Operational Performance Standards for Automatic Direction Finding (ADF) Equipment,* Radio Technical Commission for Aeronautics (RTCA), Document No. RTCA/DO-179, May 1982, Section 2.

2. "IRE Standards on Navigation Aids: Direction Finder Measurements, 1959," *Proc. IRE,* Vol. 47, No. 8, August 1959, pp. 1351–1371. Reprinted as IRE Standard 59 IRE 12.S1.

3. Terman, F.E. and J.M. Pettit, *Electronic Measurement,* New York: McGraw-Hill Book Co., 1952, pp. 405–406.

4. Haber, F., "Generation of Standard Fields in Shielded Enclosures," *Proc. IRE,* Vol. 42, No. 11, November 1954, pp. 1693–1698.

5. Sinclair, G., "Theory of Models of Electromagnetic Systems," *Proc. IRE,* Vol. 36, No. 11, November 1948, pp. 1364–1370.

6. Harrington, James B., "DF System Calibration and Correction Techniques," Watkins-Johnson Co., Palo Alto, CA, *Tech-Notes,* Vol. 9, No. 2, March/April 1982.

7. Harrington, James B., "Improving System and Environmental DF Accuracy," Watkins-Johnson Co., Palo Alto, CA, *Tech-Notes,* Vol. 9, No. 1, January/February 1982.

8. "Standards Report on Measuring Signal Strength in Radio Wave Propagation," *IEEE Standard 291-1969.*

9. "Standard Methods for Measuring Electromagnetic Field Strength for Frequencies Below 1000 MHz in Radio Wave Propagation," *IEEE Standard 302-1969.*

10. Barker, G.E., "Measurement of the Radiation Patterns of Full-Scale HF and VHF Antennas," *IEEE Trans. on Antennas and Propagation,* Vol. AP-21, No. 4, July 1973, pp. 538–544.

Chapter 10
SPREAD-SPECTRUM DIRECTION FINDING

Spread-spectrum modulation is any group of modulation formats in which an RF bandwidth much wider than necessary is used to transmit an information signal so that a signal-to-noise improvement is attained [1–6]. The most widely used spread-spectrum modulations are frequency hopping, pseudonoise direct sequence, and hybrids of the two. From a small-aperture DF viewpoint, frequency hopping is the most important because it is the most commonly used spread-spectrum modulation in the MF through UHF bands. Frequency hopping is a type of spread-spectrum modulation in which the broadband RF signal is created by hopping from one frequency to another over a large number of frequency channels. The frequencies are selected pseudorandomly in time by a pseudorandom code sequence. A frequency hopper generally hops randomly in its assigned frequency band with a constant hop duration or hop dwell time.

For direction-finding purposes, the following frequency-hopper parameters are pertinent. The total number of frequency-hop channels is F_h. The channels are generally contiguous. For example, in the 30- to 76-MHz band, a frequency hopper with 25-kHz channelization will have 1840 potential frequency-hop positions. The duration or dwell time of a hop transmission on a single-frequency channel is T_h. Clearly, if the hopper transmits for a total time T_t, the hopper performs T_t/T_h hops. Dwell times are determined by the number of hops per second, which is determined by numerous technical and operational factors. A typical hop rate for tactical military systems is 100 hops per second.

Instantaneous DF systems are mandatory for efficient frequency-hop direction finding. A combination of spatial and frequency scanning is unacceptable because it reduces the probability of efficient DF to an unsatisfactory level. The Adcock–Watson-Watt DF technique (discussed in Section 5.3.6 and diagrammed in Figure 5.36) when combined with a fast-scanning superheterodyne receiver is proving to be an efficient DF system against frequency hoppers. The pertinent parameters are (1) the time T_d required for the receiver to tune to a channel and detect the presence of a hopper and (2) the time T_m after detection required to obtain measurements for accurate angle-of-arrival information.

The probability of direction finding P_{df} on a single frequency hop is given by Deane [7] as the product of three independent probabilities as follows:

P_1 = the probability that the hopped transmission is within the frequency search band;

P_2 = the probability that the scanning DF receiver tunes to the hop channel while the hopper is still resident on the channel;

P_3 = the probability that the DF has time to detect and measure the hopper parameters before the hopper leaves assuming that $T_h > T_d + T_m$.

The probability of direction finding on a single hop is given by

$$P_{df1} = P_1 P_2 P_3 \tag{10.1}$$

where P_1 is simply the ratio of the common channels F_c between the DF search receiver and the hopper; the ratio is $P_1 = C_c/C_h$, where C_c represents the common channels and C_h the hopper channels; P_2 is based on the assumption that the time duration of a single hop T_h is much less than the total search receiver scan time T_s where

$$T_s = T_d C_s \tag{10.2}$$

and C_s is the number of search channels. In this case,

$$P_2 = T_h/(T_d C_s) \tag{10.3}$$

and P_3 is given as

$$P_3 = [1 - (T_d + T_m)/T_h] \tag{10.4}$$

Therefore, the product $P_{df1} = P_1 P_2 P_3$ is given by

$$P_{df1} = (C_c)(T_h - T_d - T_m)/(C_h C_s T_d) \tag{10.5}$$

where P_{df1} is the probability of obtaining DF data on a single hop of the frequency hopper. However, the frequency hopper transmits for a total time T_t during which time $N = T_t/T_h$ hops are made. The probability of direction finding on a frequency hopper using N hops is given by

$$P_{dfN} = 1 - (1 - P_{df1})^{T_t/T_h} \tag{10.6}$$

The time required for DF based on specified probabilities may be calculated using the expression

$$M = \ln(1 - P_{dfM})/\ln(1 - P_{df1}) \tag{10.7}$$

where M is the number of hops to obtain the specified probability values P_{dfM} and P_{df1}. Knowing M, the time to DF is given by

$$T_{dfM} = M(T_s + t_{rd}) \tag{10.8}$$

where

T_{dfM} = the time-to-DF;
T_s = the search receiver scan time [Eq. (10.2)];
t_{rd} = the average receiver delay between scans.

A major parameter is T_t, the transmission time or message length. Table 10.1 lists representative transmission durations for tactical military communications. The shortest representative transmission duration of 2 s is for AM and FM voice, which are very common modulations for tactical situations. Therefore, a T_t value of 2 s is a good choice for DF analyses.

Table 10.1
Representative Transmission Durations

Band	Modulation	Transmission Duration Typical
HF	A1A (keyed on/off)	5 s–minutes
	A2A (tones keyed on/off)	10 s–minutes
	A3E (AM/DSB)	5 s–minutes
	J3E (voice/SSB)	5 s–minutes
	F1B (FSK)	30 s–minutes
VHF-UHF	F3E (FM voice)	2 s–60 s
	A3E (AM/voice)	2 s–60 s
	F3D (FDM/MC)	hours–continuous
	PON (pulse radar)	minutes–continuous

Key
AM: amplitude modulation
DSB: double sideband
FDM: frequency division multiplex
FM: frequency modulation
FSK: frequency shift keying
MC: multichannel
SSB: single sideband

We often assume that the upper and lower bounds of frequency hopping are known *a priori* and that the search channels coincide with the hopper channels such that $C_s \equiv C_h \equiv C_c$. For military and paramilitary VHF and UHF systems, the hop frequency extent may be 100 MHz; however, a more reasonable range is 10 to 20 MHz. Using a 10-MHz hop band and 25-kHz channel spacing, the number of channels to be scanned is 400. The remaining parameters, detection time, T_d, measurement time, T_m, and scan latency time, t_{rd}, are determined by the capabilities of the DF system. Representative values for military and paramilitary VHF DF systems are $T_d = 0.6$ ms, $T_m = 0.4$ ms, and t_{rd} is generally of the order of a few microseconds depending on the method of implementation. To summarize, representative parameters are as follows:

$$C_c/C_h = 1$$
$$C_s = 400 \text{ (e.g., 10-MHz hop band and 25-kHz channels)}$$
$$T_d = 0.6 \text{ ms}$$
$$T_m = 0.4 \text{ ms}$$
$$T_h = 10.0 \text{ ms}$$
$$T_t = 2.0 \text{ s}$$
$$T_t/T_h = 200$$
$$t_{rd} = \approx 1 \text{ } \mu s$$

The use of these parameters in Eq. (10.6) returns a value of 0.9995 for the probability of DF on 200 hops. In practice, a measure of DF probability must be based on numerous DF intercepts and measurements during a transmission interval. For example, if 10 measurements are required, the effective T_t is $T_t/10$ or 0.2 s if T_t is 2 s. In this case, $T_t/T_h = 20$. Using 20 for parameter T_t/T_h in Eq. (10.6), along with the other representative parameters, returns a value of 0.534 for the probability of DF.

From a DF viewpoint, the key parameters are T_d and T_m, and, as the example demonstrates, millisecond response time ($T_d + T_m$) is required to attain acceptable performance. Clearly, instantaneous DF is required for direction finding on spread-spectrum transmissions.

An exemplary instantaneous tactical DF system is the PA2000 system by Rohde and Schwarz, Inc. Based on the Adcock–Watson-Watt technique and the use of a fast-scanning superheterodyne receiver, the PA2000 system can take bearings on frequency-hopping signals with a dwell time of about 1 ms or hop rate to 1000. The operating frequency bands are 2 to 30 MHz, 30 to 174 MHz, and 174 to 512 MHz. The HF antenna consists of an eight-element Adcock array formed by eight crossed-loop elements arrayed on a 10-m-diam circle plus an additional crossed loop acting as the sense antenna located at the phase center of the array. Above 30 MHz, the eight-element Adcock arrays use vertical dipoles. At VHF and UHF, a three-bay, mast-mounted, stacked-array configuration is used.

Cited bearing accuracy is 2° rms for short-duration frequency-hopping signals, and 1° rms for 100-ms fixed-frequency signals. Rated field strength sensitivities for 2° rms bearing accuracy are as follows:

HF: 10 μV/m;
VHF: 6 μV/m;
UHF: 6 to 20 μV/m.

Display capabilities are a frequency *versus* bearing rectangular display, a spectral display, and the conventional Watson-Watt *xy* CRT display. In the frequency-hop DF mode, the frequency *versus* bearing display provides a view of the buildup of a line of hits or DF measurements at different frequencies on the bearing of the subject hopper.

The Adcock–Watson-Watt DF technique is also well suited for DF on burst communications, which are short-duration, high-speed messages transmitted on a single frequency [1].

REFERENCES

1. Torrieri, D.J., *Principles of Secure Communications Systems,* Norwood, MA: Artech House, 1985.
2. Dillard, R.A., and G.M. Dillard, *Detectability of Spread-Spectrum Signals,* Norwood, MA: Artech House, 1989.
3. Dixon, R.C., ed., *Spread Spectrum Techniques,* New York: IEEE Press, 1976.
4. Dixon, R.C., *Spread Spectrum Systems,* 2nd Ed., New York: John Wiley and Sons, 1984.
5. Simon, M.K., *et al., Spread Spectrum Communications,* Vols. I–III. Rockville, MD: Computer Science Press, 1985.
6. Scholz, R.A., "The Origins of Spread Spectrum Communications," *IEEE Trans. Comm.,* Vol. COM-30, May 1982, pp. 822–854.
7. Deane, P. "Interception and Location of Frequency Hopping Radios," *J. Electron. Defense,* February 1987, pp. 53–56.

Appendix
METHOD OF MOMENTS COMPUTER PROGRAMS

The method of moments is a computer-based analysis technique for determining the radiation patterns of antennas in free space and on or near conducting bodies and surfaces. For most DF applications, the radiation patterns are identical to the response patterns; therefore, for analysis purposes, DF antennas may be modeled as radiators.

The conducting bodies and surfaces are modeled as either a wire-grid model or as a surface-patch model. If N wires or patches are used in the model, N discrete currents will exist on the modeled surface. The MOM algorithm is based on the numerical solution of integral equations for the currents induced on the modeled surface by an incident field or excitation source. The radiated field is computed using the integrated effects of all the N complex currents on the N wires or patches. The currents are calculated using the inverse of the $N \times N$ complex impedance matrix and the excitation source.

Moment methods analysis techniques have been extensively documented. Major literature sources are the *IEEE Transactions on Antennas and Propagation* and *Electromagnetic Compatibility.* Hansen [1] presents a collection of user-oriented moment methods papers from both IEEE and non-IEEE resources.

The moment methods approach is most efficient for electrically small to moderately sized geometries. Numerous MOM computer programs have been generated for both wire-grid and surface-patch models. The wire-grid modeling approach is particularly effective for predicting far-field patterns, which are generally of most interest for DF applications.

The purpose of this appendix is to highlight representative MOM computer programs applicable to small-aperture DF analysis. Mainframe-, minicomputer-, and microcomputer-based programs exist.

Mainframe MOM Programs

Two major mainframe-based MOM programs are GEMACS (General Electromag-netic Analysis of Complex Systems) and NEC (Numerical Electromagnetic Code).

GEMACS [2, 3] is used in the analysis of electrically large host platforms such as one that has tens of square wavelengths as a two-dimensional surface or hundreds of wavelengths for one dimension. GEMACS can accommodate several thousand unknowns for both radiation and scattering situations. Major attributes of GEMACS are out-of-core manipulations and the banded matrix iteration (BMI) technique. These attributes make the moment methods analysis of large systems practical and cost effective. A basic strength of GEMACS is the storage and processing of large quantities of data.

The NEC program [4] is similar to GEMACS in that the NEC program deals with multiwavelength platforms and accommodates several thousand unknowns. NEC uses both an electric-field integral equation (EFIE) and a magnetic-field integral equation (MFIE). NEC provides modeling flexibility in that both integral solutions may be combined. NEC computation time is reduced by exploiting rotational or planar symmetries in the host platform. Realistic, imperfect ground conditions are accommodated by the use of the Sommerfield integral [1, p. 325]. The excitation sources may be an applied Hertzian source or an incident plane wave with linear or elliptical polarization. Nonradiating, lumped-constant networks and transmission lines may be included in the NEC analysis.

Typical *inputs* to the mainframe-based MOM programs are as follows:

- Geometric wire or patch descriptions of the host platform or counterpoise;
- Excitation points and levels;
- Lumped or distributed complex impedance loading;
- Ground conductivity and dielectric parameters;
- Nonradiating elements such as coupling networks and transmission lines.

Typical *outputs* of the mainframe-based MOM programs are as follows:

- Near- and far-field, E- and H-field, azimuth and elevation radiation parameters *versus* spatial aspect angles as a function of frequency, polarization, and ground constants;
- Complex current distributions on the wires and patches;
- Complex input impedance and power at the excitation points.

Minicomputer MOM Programs

Numerous minicomputer-based MOM programs have been generated and documented. The *IEEE Transactions on Antennas and Propagation, Electromagnetic*

Compatibility, and Hansen [1] document many of these programs. Allen and Ryan [5] describe programs especially for MOM analysis of MF and HF antennas on aircraft. A large number of the minicomputer-based programs are specialized for specific configurations. For example, bodies-of-revolution (BOR) programs exist for host platforms with curved symmetrical geometry. Also, programs based on "stick" models of the host platform exist for LF, MF, and lower HF applications.

Compared to mainframe-based programs, minicomputer-based programs have limited capabilities. A typical capacity of several hundred unknowns requires a relatively simple configuration and a limited upper frequency limit. Further, in general, imperfect conductors and lossy ground cannot be handled easily by minicomputer MOM programs. Storage and processing of large quantities of data are limited. Generation of the wire-grid and surface-patch models requires manual efforts.

In spite of significant disadvantages, minicomputer MOM programs are very useful for comparative evaluations and relatively uncomplicated situations. For example, minicomputer-based MOM programs are capable of analyzing small-aperture DF antennas on host platforms such as jeeps, small helicopters, and light aircraft through the HF band. Some programs are capable of extending the analysis into the VHF band. A representative minicomputer-based MOM program, which is well-validated and extensively-applied, is the WR-SYR (WIRE-SYRACUSE) program [6, 7].

Microcomputer MOM Programs

A reduced version of the NEC mainframe program has been formulated by Julian, Logan, and Rockway [8] for use on microcomputers. Called MININEC, versions 1 and 2 of the program are written in BASIC [8]. Later versions have been written in FORTRAN [9, 10]. A major advantage of the MININEC programs is the use of a compact code based on an algorithm developed by Wiltron [1, p. 58] for an efficient numerical solution of the electric field integral equation. MININEC programs, which use wire-grid models, can handle about 50 unknowns. MININEC programs solve for (1) far-field patterns in free space and over a flat, perfectly conducting, infinite ground plane and (2) complex impedance and current on the wires. MININEC accommodates lumped parameter complex impedance loading of the wires.

PC-based MININEC programs are commercially available both with [11] and without [12, 13] accompanying tutorial and instructional material. Another PC-based MOM program is the Wire Antenna and Scatterer Analysis program [14]. Note that PC-based MOM programs are, by necessity, capability limited for DF application analyses.

REFERENCES

1. Hansen, R.C., ed., *Moment Methods in Antennas and Scattering,* Norwood, MA: Artech House, 1990.
2. Balestri, R.J., *et al.,* "General Electromagnetics Model for the Analysis of Complex Systems-User's Manual," Final Technical Report RADC-TR-77-137, Contract F30602-74-C-0182, Vol. I (of two), ADA040026, Rome Air Development Center, April 1977.
3. Balestri, R.J., *et al.,* "General Electromagnetics Model for the Analysis of Complex Systems—Engineering Manual," Final Technical Report RADC-TR-77-137, Vol. II (of two), ADA040027, Rome Air Development Center, April 1977.
4. Burke, G.L., and A.J. Poggio, "Numerical Electromagnetic Code—Method of Moments," Technical Document No. 116, Naval Electronics Systems Command (ELEX 3041), 18 July 1977, Revised 2 January 1980. Consists of three parts. Part I: Program Description—Theory; Part II: Program Description—Code; and Part III: User's Guide.
5. Allen, W.P. Jr., and C.E. Ryan, Jr., "Aircraft Antennas," Chapter 37 in *Antenna Engineering Handbook,* Johnson, R.C., and H. Jasik, eds., NY: McGraw-Hill Book, Co., 1984, Section 37-4.
6. Kuo, D.C., and B.J. Strait, "Improved Programs for Analysis of Radiation and Scattering by Configurations of Arbitrarily Bent Thin Wires," Scientific Report 15, Contract F19628-68-C-0180, AFCRL-72-0051, Syracuse, New York: Syracuse University, January 1972.
7. Kuo, D.C., *et al.,* "Analysis of Radiation and Scattering by Arbitrary Configurations of Thin Wires," *IEEE Trans. Antennas Propag.,* Vol. AP-20, November 1972, pp. 814–815.
8. Julian, A.J., J.C. Logan, and J.W. Rockway, "MININEC: A Mini-Numerical Electromagnetics Code," Technical Document 516, Naval Ocean Systems Center, September 1982, ADA121535.
9. Logan, J.C., and J.W. Rockway, "The New MININEC (Version 3): A Mini-Numerical Electromagnetic Code," Technical Document 938, Naval Ocean Systems Center, September 1986.
10. Carr, A., "A FORTRAN Update of the MININEC Electromagnetics Modeling Program," *RF Des.,* February 1990, pp. 59–60.
11. Li, S.T., J.C. Logan, J.W. Rockway, and D.W.S. Tam, *The MININEC SYSTEM: Microcomputer Analysis of Wire Antennas,* Norwood, MA: Artech House, 1988.
12. Lewallen, R., "Antenna Analysis Program," *RF Des.,* February 1990, p. 63.
13. "RF Design Software Service," *RF Des.,* February 1990, p. 65.
14. Djordjevic, A.R., *et al., Wire Antenna and Scatterer Analysis,* Norwood, MA: Artech House, May 1990.

Glossary

Acronym or Abbreviation	Definition
ADF	Automatic direction finder
ADDF	Automatic digital direction finder
AGC	Automatic gain control
AM	Amplitude modulation
AMP	Amplifier
AOA	Angle of arrival
AZ/EL DF	Azimuth–elevation direction finder
	Azimuth–elevation direction finding
BFO	Beat frequency oscillator
BW	Bandwidth
CEP	Circular error probability
CNR	Carrier-to-noise ratio
COMINT	Communication intelligence
CRT	Cathode-ray tube
CW	Continuous wave
DF	Direction finder, direction finding
DFT	Discrete Fourier transform
DOA	Direction of arrival
DS	Direct sequence
ELT	Emergency locator transmitter
EPIRB	Emergency position indicating radiobeacon
ESM	Electronic support measures
FAGC	Fast-acting automatic gain control
FFT	Fast Fourier transform
FH	Frequency hop, hopping, or hopper
FLOT	Forward line of troops
FM	Frequency modulation
FOT	*See* OWF
FOV	Field of view
GCB	Great-circle bearing
GDOP	Geometric dilution of precision
GPS	Global positioning (satellite) system
HF	High frequency

Acronym or Abbreviation	Definition
ICM	Interrupted carrier modulation
IF	Intermediate frequency
LCD	Liquid crystal display
LED	Light-emitting diode
LF	Low frequency
LOB	Line of bearing
LOP	Line of position
LORAN	Long-range navigation
LUF	Lowest usable frequency
MCW	Modulated continuous wave
MF	Medium frequency
MLE	Maximum likelihood estimator
MOM	Method of moments
MUF	Maximum usable frequency
NCS	Net control station
NDB	Nondirectional beacon
NEL	Noise equivalent level
NF	Noise figure
NOE	Nap of the earth
NVIS	Near-vertical incidence signaling
OTH	Over the horizon
OWF	Optimum working frequency (FOT)
PCA	Polar cap absorption
PM	Phase modulation
PN	Pseudonoise
RCVR	Receiver
RDF	Radio direction finder, radio direction finding
RF	Radio frequency
SATNAV	Satellite navigation
SD	Standard deviation
SEP	Spherical error probability
SID	Sudden ionospheric disturbance
SNR	Signal-to-noise ratio
SSB	Single-sideband
SSL	Single station location
T	Time
TACAN	Tactical air navigation
TDOA	Time difference of arrival
TID	Traveling ionospheric disturbance
TX	Transmitter
UHF	Ultra-high frequency
VHF	Very high frequency
VLF	Very low frequency

INDEX

The Artech House Radar Library

David K. Barton, *Series Editor*

Modern Radar System Analysis Software and User's Manual by David K. Barton and William F. Barton

Monopulse Principles and Techniques by Samuel M. Sherman

Monopulse Radar by A.I. Leonov and K.I. Fomichev

MTI and Pulsed Doppler Radar by D. Curtis Schleher

Multifunction Array Radar Design by Dale R. Billetter

Multisensor Data Fusion by Edward L. Waltz and James Llinas

Multiple-Target Tracking with Radar Applications by Samuel S. Blackman

Multitarget-Multisensor Tracking: Advanced Applications, Yaakov Bar-Shalom, ed.

Over-The-Horizon Radar by A.A. Kolosov, et al.

Principles and Applications of Millimeter-Wave Radar, Charles E. Brown and Nicholas C. Currie, eds.

Principles of Modern Radar Systems by Michel H. Carpentier

Pulse Train Analysis Using Personal Computers by Richard G. Wiley and Michael B. Szymanski

Radar and the Atmosphere by Alfred J. Bogush, Jr.

Radar Anti-Jamming Techniques by M.V. Maksimov, *et al.*

Radar Cross Section by Eugene F. Knott, *et al.*

Radar Detection by J.V. DiFranco and W.L. Rubin

Radar Electronic Countermeasures System Design by Richard J. Wiegand

Radar Evaluation Handbook by David K. Barton, *et al.*

Radar Evaluation Software by David K. Barton and William F. Barton

Radar Propagation at Low Altitudes by M.L. Meeks

Radar Range-Performance Analysis by Lamont V. Blake

Radar Reflectivity Measurement: Techniques and Applications, Nicholas C. Currie, ed.

Radar Reflectivity of Land and Sea by Maurice W. Long

Radar System Design and Analysis by S.A. Hovanessian

Radar Technology, Eli Brookner, ed.

Receiving Systems Design by Stephen J. Erst

Radar Vulnerability to Jamming by Robert N. Lothes, Michael B. Szymanski, and Richard G. Wiley

RGCALC: Radar Range Detection Software and User's Manual by John E. Fielding and Gary D. Reynolds

SACALC: Signal Analysis Software and User's Guide by William T. Hardy

Secondary Surveillance Radar by Michael C. Stevens

SIGCLUT: Surface and Volumetric Clutter-to-Noise, Jammer and Target Signal-to-Noise Radar Calculation Software and User's Manual by William A. Skillman

Signal Theory and Random Processes by Harry Urkowitz

Solid-State Radar Transmitters by Edward D. Ostroff, *et al.*

Space-Based Radar Handbook, Leopold J. Cantafio, ed.

Spaceborne Weather Radar by Robert M. Meneghini and Toshiaki Kozu

Statistical Theory of Extended Radar Targets by R.V. Ostrovityanov and F.A. Basalov

The Scattering of Electromagnetic Waves from Rough Surfaces by Peter Beckmann and Andre Spizzichino

VCCALC: Vertical Coverage Plotting Software and User's Manual by John E. Fielding and Gary D. Reynolds

www.ingramcontent.com/pod-product-compliance
Lightning Source LLC
Chambersburg PA
CBHW021430180326
41458CB00001B/202